C. L. Aurelio
March 2000

Robust Modulation Methods and Smart Antennas in Wireless Communications

Bruno Pattan
Senior Member of the Technical Staff
Federal Communications Commission
Office of Engineering and Technology

Prentice Hall PTR
Upper Saddle River, NJ 07458
http://www.phptr.com

ISBN 0-13-022029-9

Library of Congress Cataloging-in-Publication Data

```
Pattan, Bruno
     Robust modulation methods and smart antennas in wireless
   communications / by Bruno Pattan.
       p.  cm.
     Includes bibliographical references and index.
     ISBN 0-13-022029-9
     1. Adaptive antennas.  2. Wireless communication systems.
   3. Modulation (Electronics)  4. Spread spectrum communications.
   I. Title.
   TK7871.6.P38  1999
   621.382'4--DC21                                        99-11799
                                                              CIP
```

Editorial/production supervision: *Joan L. McNamara, Patti Guerrieri*
Cover designer: *Anthony Gemmellaro*
Cover design director: *Jerry Votta*
Manufacturing manager: *Alan Fischer*
Marketing manager: *Lisa Konzelmann*
Acquisitions editor: *Bernard Goodwin*
Editorial assistant: *Diane Spina*
Book design and layout: *Aurelia Scharnhorst*

© 2000 by Prentice Hall PTR
  Prentice Hall, Inc.
  Upper Saddle River, New Jersey 07458

Prentice Hall books are widely used by corporations and government agencies for training, marketing, and resale.
The publisher offers discounts on this book when ordered in bulk quantities.
For more information, contact Corporate Sales Department. Phone: 800-382-3419;
Fax: 201-236-7141; E-mail: corpsales@prenhall.com
Or write: Prentice Hall PTR, Corp. Sales Dept., One Lake Street, Upper Saddle River, NJ 07458

The views or conclusions contained in this book are those of the author and should not be interpreted as necessarily representing the official policies, either expressed or implied of the Federal Communications Commission.

Sprint and the diamond logo are registered trademarks of Sprint Communications Company L.P., used uder license. Sprint PCS is a service mark of Sprint Communications Company L.P.Product names mentioned herein are the trademarks or registered trademarks of their respective owners.

All rights reserved. No part of this book may be reproduced, in any form or by any means, without permission in writing from the publisher.

Printed in the United States of America
   10 9 8 7 6 5 4 3 2 1

ISBN 0-13-022029-9

Prentice-Hall International (UK) Limited, *London*
Prentice-Hall of Australia Pty. Limited, *Sydney*
Prentice-Hall Canada Inc., *Toronto*
Prentice-Hall Hispanoamericana, S.A., *Mexico*
Prentice-Hall of India Private Limited, *New Delhi*
Prentice-Hall of Japan, Inc., *Tokyo*
Prentice-Hall (Singapore) Pte. Ltd., *Singapore*
Editora Prentice-Hall do Brasil, Ltda., *Rio de Janeiro*

This book is dedicated to my mother and father, who never went to school, but realized the benefit of education.

Table of Contents

Preface xi
Acknowledgments xii

1 In Pursuit of Bandwidth Efficiencies for Wireless Terrestrial and Satellite Communications 1

1.1 Introduction 1

2 Bandwidth-Efficient Modulation Techniques 5

2.1 Introduction 5
2.2 Bandwidth and Power Efficiency Plane 6
2.3 Conclusions 11
2.4 References 13

3 Higher Order Modulation Methods 15

3.1 Introduction 15
3.2 Signal State Space Diagrams 21
3.3 Performance Representations 23
3.4 Conclusions 29
3.5 Glossary of Terms 31
3.6 References 32

4 Dynamics of Linear and Continuous Phase Modulation Methods in Digital Communications 33

4.1 Introduction 33
4.2 Linear Modulation 34
4.3 Continuous Phase Modulation 42
4.4 Phase Trellises in CPM 48
4.5 GMSK Modulation 51
4.6 Tamed Frequency Modulation (TFM) 58
4.7 Signal Orthogonality 59
4.8 Conclusions 64
4.9 References 65

5 Error Control Coding 69

5.1 Introduction 69
5.2 Code Families 70
5.3 Code Performance 75
5.4 Block Coding 82
5.4.1 Popular Block Codes 85
5.5 Convolutional Encoding 89
5.6 Concatenated Coding 98
5.7 Interleaving 98
5.8 Coding Break-through 102
5.9 Conclusions 102
5.10 References 104

6 Trellis Coded Modulation (Codulation) 105

6.1 Introduction 105
6.2 The Theory 107
6.3 Attributes of Trellis Coded Modulation 112
6.4 Practical Systems 117
6.5 Performance Degraders 121
6.6 Conclusions 122
6.7 Glossary of Terms 123
6.8 References 125

7 Spread Spectrum Communication Systems 127

- 7.1 Introduction 127
- 7.2 Spread Spectrum Techniques 129
  - *7.2.1 Direct Sequence Concept 131*
  - *7.2.2 Frequency Hopping Concept 133*
- 7.3 Code Generation 133
- 7.4 Codes for Spread Spectrum Multiplexing 140
- 7.5 Spread Spectrum Interference Analysis 146
  - *7.5.1 One-on-One 146*
  - *7.5.2 Multiply Access Interference Scenario 148*
- 7.6 The Multipath Phenomenon 152
  - *7.6.1 RAKEing in Performance 153*
  - *7.6.2 RAKE Receiver 154*
- 7.7 Purely Random or Pseudo-Random— What's the Difference? 155
- 7.8 Conclusions 155
- 7.9 Glossary of Terms 156
- 7.10 References 158

8 Terrestrial-based Wireless Communications 161

- 8.1 Introduction 161
- 8.2 Frequency Bands of Operation 164
- 8.3 Interference Analysis 166
- 8.4 Increasing Capacity 174
- 8.5 Cellular Standards 179
- 8.6 Personal Communications Service 185
- 8.7 Conclusions 189
- 8.8 References 190

9 The Butler Matrix 193

- 9.1 Introduction 193
- 9.2 Planar Array Beams 194
- 9.3 Multiple Volumetric Beams 197
- 9.4 Butler Array Application 202
- 9.5 Conclusions 202
- 9.6 References 202

10 Sidelobe Cancellers in Smart Antenna Applications 203

- 10.1 Introduction 203
- 10.2 Single Interferer Sidelobe Canceller 203
- 10.3 Multiple Interferers 207
- 10.4 Conclusions 207
- 10.5 References 207

11 A Look at Switched-Beam Smart Antennas 209

- 11.1 Introduction 209
- 11.2 Trunking Efficiency 214
- 11.3 Smart Antennas 214
- 11.4 Configurations 221
- 11.5 Conclusions 224
- 11.6 References 226

12 Deterministic Signals, Random Noise, and Coherent Noise (Pseudo) Combining in an Array Antenna 227

- 12.1 Introduction 227
- 12.2 Coherent Signals 229
- 12.3 Coherent Noise 232
- 12.4 An Adaptive Array in a Quiescent Signal Field 234
  - *12.4.1 An Example 237*

13 Adaptive Arrays in Cellular Communications 241

- 13.1 Introduction 241
- 13.2 The Theory 243
- 13.3 Simulation Results 248
- 13.4 Conclusions 250
- 13.5 References 250

14 Summary – Smart Antennas in Cellular Communications 253

- 14.1 Introduction 253
- 14.2 Adaptive Array Genre 256

14.3 Where Are Smart Antennas Going?   260
14.4 Conclusions   260

A Gaussian Low-Pass Filter   261

B Scattering Matrix of the Quadrature Hybrid   263

C Example of Trunking and Erlang Tables   267

D Glossary of Terms   269

Index   279

The Author   287

Preface

The aim of this book is to regale the reader with an overview of some of the technologies peculiar to wireless communications. I have addressed what I believe are important aspects of the subject. The material is tailored for technical personnel working in the field of wireless, who are seeking additional information on the technologies in this area. The practicing engineer will find the text to contain useful information concerning the design of wireless systems. The material presented is also suitable for senior undergraduate or graduate students majoring in communications. The prerequisite knowledge is a first course in communication theory, some exposure to probability and random noise theory, and a nodding acquaintance with matrices.

Wireless communications have consistently exceeded the capacity of available technology. The exponential increase in voice service (mobile in particular), together with the ever-growing demand for data services, have pushed current systems beyond their capacities. There is therefore a continuous pursuit to satisfy these burgeoning demands and for advancing the technological frontiers.

The coverage in this book is broad, encompassing subjects from signal formats to smart antennas, with the latter developing in the continuous pursuit of more capacity. The material is not rigorous, but is reader-friendly with a tutorial slant. The text is complemented with numerous figures to make the presentation more lucid.

The various technologies described in this book are as follows: The first two chapters deal with spectral efficiencies and power efficiencies within Shannon bounds. Providing wireless service in a spectrum where there is paucity of spectrum is an ever-present challenge. Chapter 3 discusses various higher order modulation methods in the presence of limited bandwidth, which can achieve increased spectral efficiency (b/s-Hz), but with a concomitant increase in power requirements. Chapter 4 deals with modulation methods which provide high spectral efficiency and robustness in a stressed environment. The latter includes fading induced amplitude fluctuations in the received signals and nonlinearities in the communications channel, and hence permits the utilization of efficient-C amplifiers. This is followed by Chapter 5, which deals with error-correcting codes with coding gain — a necessary adjunct in wireless to cope with the fading signal environment and other deleterious interference. Both random and bursty errors are generated, which are combatted by various coding schemes. Chapter 6, Trellis Coded Modulation, is a compliment to the coding chapter. This modulation type provides coding gain without sacrificing additional bandwidth and is truly a breakthrough in coding theory.

Chapter 7, Spread Spectrum Communications, describes a cellular standard now used in the U.S. This standard mitigates interference from systems using co-channel operation by tagging each channel with its unique identifying orthogonal code. Each signal channel sees the other channels as adding noise-like interference to its channel (which puts a bound on capacity). It potentially can significantly increase capacity to cellular systems, even though this has yet to be established.

Chapter 8, Terrestrial Cellular Communications, presents some of the concepts used in terrestrial cellular, including the various signal formats and performance specifications used by various standards, which have been developed and used globally.

The next few chapters deal with the evolution of smart antennas. These antenna systems use phased arrays to produce beams in space which can increase the capacity of a system. Chapter 9 starts with a discussion of the Butler matrix, which is an integral component of some smart antennas. The Sidelobe Canceller, discussed in Chapter 10 had its origin in radar and was used to reduce interference coming into radar antenna sidelobes. It has few applications in cellular, but is presented for historical value and lays the groundwork for smart arrays. Chapters 11-13 deal in more detail with the two basic types of smart arrays — that is, switched-multiple beam and adaptive array configurations. The attributes and shortcomings of both are given. The last chapter, Chapter 14, is a summary of smart antennas and where they are going in cellular communications.

All chapters are complemented by a list of references through which the reader may seek additional information.

Acknowledgments

I would like to acknowledge my stimulation conversations with Roman Zaputowycz, formerly of Bell Atlantic Mobile, on cellular systems. In addition, the profound discussions I had with Professor Irving Kalet of Israel contributed to my knowledge of the subtle points on signal theory. I have attempted to give adequate credit to all sources of the above cited material, but may have occasionally manifested sins of omission. Of course, any glitches which may have slipped through are solely "mea culpa."

Bruno Pattan
Washington, D.C.

CHAPTER 1

In Pursuit of Bandwidth Efficiencies for Wireless Terrestrial and Satellite Communications

1.1 Introduction

To the world in general, the term wireless communications implies personal telephony as manifested by handset communications, or simply radiotelephony. To others, wireless connotes different services such as paging, which displays text and numeric messages (one-way or two-way, e.g., Motorola's Pagewriter 2000). To some it means short-range cordless telephone (CT-2, DECT), and to others data transmissions, which are in the nascent stage but will increase with the use of handheld computers. Whatever wireless communications means to different people, it is anticipated that the demand for mobility will make wireless technology the primary source of voice communications in the future.

To date, the route taken to the above services has been mostly via terrestrial means, but emerging satellite-based personal communications providing voice and data are coming to fruition and will provide ubiquitous and seamless wireless connectivity globally. Witness the non-geostationary satellites (LEOs and MEOs), e.g., IRIDIUM, Globalstar, IOC, which are operational or near operational. These will provide voice, paging, fax, and messaging to fixed and mobile locations. The world's appetite for high-speed data services continues to grow and satellites appear to be the next best delivery system for high-speed access to the Internet, e.g., Teledesic (Gates, McCaw, Motorola), Skybridge (Alcatel). However, because of the limited power provided by satellites and mobile ground transceivers and a substantial building RF-altenuator, the link margins are not adequate to provide voice service directly into buildings.

From its humble beginnings in the early 1980s, using analog modulation (AMPS, IS-553, NMT, TACS) untethered personal cellular communications (voice) has experienced unprecedented growth, and has become a leviathan encompassing the entire globe. The trend now is to more efficient digital communications with added attributes above analog systems. These include voice mail, caller ID, call waiting, call forwarding, paging, and lately Email.

Wireless voice networks span a broad range which stretches from local cordless telephones to global satellite delivery systems. In the U.S. alone, there are more than 70 million subscribers and this number is growing. Worldwide, the number is close to 250 million.

Concomitant with this growth has sprouted a host of technical and regulatory problems which require solution. Clearly some are extensions of wireless technology, while others are peculiar to wireless service. For example, multi-path fading demonstrates statistical variations in signal levels, which are more hostile than AWGN (Additive White Gaussian Noise) and affect the BER. Fading also introduces bursty errors (continuous error bits), which can be handled using Reed-Solomon codes and frequently complemented by interleaving, or by employing the latter method singularly. Delay spread (frequency-selective fading), in which a bit arrives at the receiver at different times because of the different paths taken, causing bits to run into each other and thus cause inter-symbol interference (ISI), limits the usable digital signalling rate for a given error rate. This problem has been combated by equalization in the receiver. The use of spread spectrum can temper this delay effect since delay is frequency-sensitive and the wide spectrum allows most frequencies to pass unscathed.

Co-channel interference in cellular systems, where spectrum is shared by frequency reuse, is also a problem because of the proximity of the cells to each other. Here, digital systems also ameliorate this problem. In analog cellular systems, e.g., AMPS, the FM threshold criterion for C/I is about 17 dB. In digital systems, the acceptable threshold C/Is are in the range of 7-14 dB. The other source of interference is intra-system interference, resulting from inter-symbol interference. Both limit the number of users supported by a system.

On the other hand, handsets provide a host of problems, including packaging high-density circuits in a limited volume, i.e., removal of heat and poorer consumption. MMIC (Monolithic Microwave Integrated Circuit) and VLSI (Very Large Scale Integration) technologies alleviate these problems by providing adequate battery power to ensure long talk and standby times, nevertheless, this continues to be a problem. Of course, transceivers that are small and moderately priced are also important. Economies of scale in manufacturing and distribution exert a downward trend in cost. However, price in some instances is not a factor, since organizations will sell transceivers at a nominal price to have users sign up for the service.

From the early days of analog systems, the trend has been to move toward digital communications. Benefits accrue from the use of digital communications, including efficient modulation schemes, use of error correction coding, use of interleaving, increased voice activation (picks up almost a factor of two in capacity improvement), both voice and data compression, use of networks to combat multi-path fading (intra-receiver time diversity RAKE, adaptive equalization), and security.

All digital communications use low bit rate speech coding to increase system capacity, but are not necessarily accompanied by an improvement in voice fidelity (contrary to what one reads in the open literature); in fact, they may even degrade it. Toll-quality voice (Mean Operating Score) (MOS : 4.0-4.5) is hard to come by at these low bit rates. Wireless digital voice now falls in the "communication quality" range with a MOS in the range of 3.0-4.0 (MOS: mean operating score – figure of merit of voice quality).

We must also realize that the use of some of the signal processing schemes alluded to above introduces delays which may not be insignificant. Clearly, they are not as severe as those found in Geostationary Satellite Orbit (GSO) satellite transmissions (>250 ms), but nevertheless they could be a problem, especially when there is a concatenation of networks. Some of these delay problems associated with satellite transmissions have been avoided by going to wireline fiber and other *quasi*-terrestrial means (low-orbit communications satellites (IRIDIUM, Globalstar, Intermediate Orbit Communications (IOC)).

The advent of digital has forced researchers to use different types of signal formatting, which are incompatible with each other. In the U.S., there are three interim standards[1] that appear to predominate: IS-54 (136) or D-AMPS, IS-95 (CDMA), and GSM (in PCS-1900). A popular modulation scheme includes the GSM constant envelope GMSK (Gaussian Minimum Shift Keying), which has a spectral efficiency of 1.3 b/s-Hz. This is an elegant signal format which is insensitive to received signal amplitude fluctuations. It also allows the use of efficient Class-C amplifiers without fear of spectral regrowth. On the other hand, where there is a paucity of spectrum, spectral efficiency prevails with IS-54, which uses $\pi/4$-DQPSK (with a 35% filter roll-off factor, Nyquist), and which has a theoretical efficiency of 2 b/s-Hz (practically 1.6 b/s-Hz). This modulation normally requires the use of less efficient Class-A amplifiers (to prevent spectrum regeneration), adding more weight to the handset. Constant envelope

---

1. The U.S. (FCC) has not established a digital cellular standard.

modulation systems are generally less efficient than linear modulation systems like $\pi/4$-DQPSK. The CDMA (Code Division Multiple Access) standard IS-95 uses direct sequence spread spectrum with QPSK/BPSK modulation. It has a spectral efficiency of 0.98.

All proponents of the above standards claim to provide greater capacity than the U.S. entrenched FM-AMPS system. Claims of up to 20 times greater improvement have been bandied about. And, claims of near toll quality voice fidelity are suspicious. It is realistic to expect that both of these claims are inflated.

It is unfortunate that these systems are not mutually compatible, which in turn suggests that handsets must be multi-mode types to operate and enable roaming different parts of the country and the world. Europe has basically agreed on the digital GSK standard. Therefore, roaming is possible in most of Europe.

Significant capacity improvement results have been realized with cellular systems. A capacity increase has also been realized by cell sectorization (3-6), which is facilitated by the use of directional antenna beams.

A more recent emerging technology that is showing considerable promise is the smart antenna system. As stated by Dr. A. Viterbi, "Spatial processing remains the most promising if not the last frontier in the evolution of multiple access systems." Essentially what he is saying is that dynamic space diversity, or space division multiple access (SDMA), is necessary to further increase capacity, above using classical means, by increasing C/I and reducing ICI. This results in smaller cell clustering ($N<7$). This area includes switched multiple beams and adaptive arrays of the Widrow genre. In smart antenna systems, the basic building blocks are array structures consisting of multiple antenna elements. Their application is now limited to fixed base stations where "real estate" is available, but not at the user transceiver level. These concepts, and others indicated above, will be discussed in subsequent sections of this book.

CHAPTER 2

Bandwidth-Efficient Modulation Techniques

2.1 Introduction

Shannon, in Theorem 17 of his classical paper [2], states that the "capacity of a channel of bandwidth B perturbed by white thermal noise of power N when the average transmitter power is limited to P is given by the following: It is not possible to transmit at a higher bit rate by any encoding system without a definite positive frequency of errors." Shannon's theorem establishes the maximum rate (I_{max}) at which information can be reliably transmitted through a channel, but does not indicate how to construct practical codes and/or decoding schemes that will achieve this capacity. The practical problem of achieving anything like the performance promised by Shannon is extremely difficult. This is where the challenge lies for communication theorists.

$$\text{Information rate}: I_{max} = B \log_2(N+S)/N \quad \text{bits/second (bps)} \qquad (2.1)$$

With the logarithm taken to base 2, the capacity is in bits per second. This is the maximum number of bits that can reliably be sent over a channel.

The channel capacity is given as

$$C = I/B = \log_2(N+S)/N = \log_{10}(1 + (S/N)/\log_{10} 2 \quad \text{b/s-Hz}^1 \qquad (2.2)$$

where B is channel bandwidth (Hz), S is average signal received power (W) bounded by B, and N is additive white Gaussian noise (AWGN), (W). Note that $N = kTB$, where $kT = N_o$ = noise spectral density.

---

1. $\log_2 X = \log_{10} X / \log_{10} 2 = 3.32 \log_{10} X$

This gives the upper limit that can be reached in the way of reliable data communications over a Gaussian channel. Notice that you can exchange the bandwidth (B) for the signal-to-noise ratio (S/N). For example, assume you want to transmit data at a rate of 9000 bps and the channel has a bandwidth of $B = 3$ kHz. It is clear that to transmit the data at a rate of 9 kbps, the channel capacity has to be at least 9 kbps. If the capacity is less than this, errorless transmission is not possible. Or, in symbolic form from above, the rate, R, is

$$R < C = B\log[1 + (S/N_oB)] \quad \text{bps} \tag{2.3}$$

With a channel bandwidth of 3 kHz and a required channel capacity of at least 9 kbps, the minimum required S/N is therefore (finding the anti-log)

$$S/N = 2^{C/B} - 1 = 2^{9/3} - 1 = 7 \tag{2.4}$$

If the channel bandwidth were 9 kHz, the required $S/N = 1$.

We also notice the signal power, S, times the bit interval, T, is equal to the energy per bit. Therefore rewriting Equation (2.3) we obtain

$$R < C = B\log[1 + (E_bR_b/N_oB)] \tag{2.5}$$

Therefore, to "push" the bit rate R_b (to 9 kbps, say), R_b/N_o, E_b/N_o must satisfy the following relationship from Equation (2.5):

$$E_b/N_o = [1/(R_b/B)](2^{R_b/B} - 1)$$
$$\text{or } E_b/N_o = 10\log[(B/R_b)(2^{R_b/B} - 1)] \quad \text{dB} \tag{2.6}$$

where R_b/B is in b/s-Hz.

This is not unlike Equation (2.4) where we replaced S/N (Equation (2.3)) by $(E_b/R_b)/N_oB$ to obtain Equation (2.6). Equation (2.6) determines what is theoretically possible for a band-limited channel.

2.2 Bandwidth and Power Efficiency Plane

Equation (2.6) is plotted in Figure 2–1 [3]. The ordinate reflects the amount of data that can be transmitted in a specified bandwidth within a given time.[2] The Shannon bound occurs when B approaches ∞ ($R/B \to 0$). By Expanding $2^{R_b/B}$ in series form (from Burington's handbook) [6] we obtain

---

2. It is interesting to observe on a family of "water fall" (E_b/N_o) curves, that the curves become more precipitous (with increasing bandwidth) as you move to the left and eventually become vertical as the bound is reached (if soft decision detection is used).

Bandwidth and Power Efficiency Plane

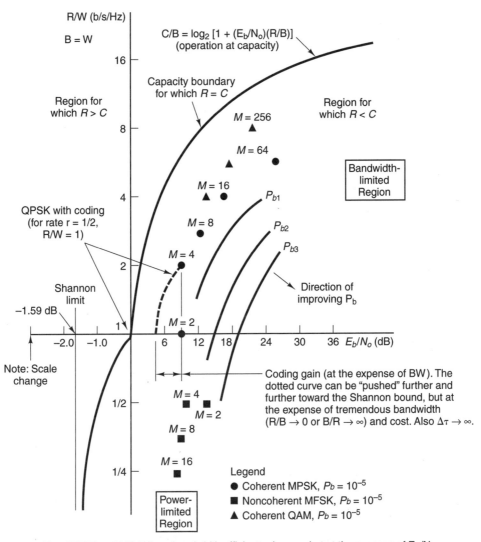

Figure 2–1 Comparison of bandwidth/power efficiency of different modulation types. Copyright © 1988 by Prentice-Hall, Inc. Reprinted with the permission of the publisher.

$$E_b / N_o = 10\log_{10}(B / R_b) \, [1 + (R / B)\ln 2 + (1/2 \, (R / B)\ln 2) + \ldots - 1$$
$$\text{Limit } (B / R_b) \to \infty \, (E_b / N_o) = \log_2 2 = 0.693 = -1.59 \text{ db-Hz} \quad (2.7)$$

Therefore, a zero-error transmission can in principle be accomplished at $E_b/N_o = -1.59$ dB when noise is additive white Gaussian (AWGN). Below $E_b/N_o = -1.59$ dB there can be no error-free performance, even if B is arbitrarily large or if R is vanishingly small. Therefore, and actual signalling and receiver method will be to the right and below the zero-error solid curve.

Actually, practical systems operate well above this curve. It is not possible to reach the Shannon bound because the bandwidth and delay become infinite. Operating close to this value is therefore challenging and complex.

Superimposed are the operating points for MPSK, MFSK, and MQAM for bit error rates (BERs) equal to 10^{-5}. Also, notice that the ordinate of the plotted points gives the bandwidth efficiency, $R/B = \log M$ b/s-Hz. For systems which are *bandwidth-limited* (but with adequate E_b/N_o), it is desirable to utilize modulation techniques that allow the transmission of more bits per second (movement to the left), or higher bandwidth efficiency, R/B.

Notice that for MPSK, increasing M (alphabet size) increases the bandwidth efficiencies. However, for 2PSK and 4PSK (QPSK), the E_b/N_o is the same even though 4PSK has twice the bandwidth efficiency. QPSK may be considered as two BPSK signals orthogonal to one another. BPSK and QPSK have the same E_b/N_o, but differ by 3 dB in the signal-to-noise ratio (BPSK: $E_b/N_o = S/N$, QPSK: $E_b/N_o = (S/N) - 3$ dB).

If we have a situation where energy is scarce for the desired bit error rate (P), we move to the right and down, lowering the R/B to meet the P requirement (as indicated by the three partial solid curves in Figure 2–1). For MFSK, on the other hand, increasing M reduces the bandwidth efficiency, but reduces the E_b/N_o requirement. M-ary FSK systems trade bandwidth efficiency for power. As M approaches infinity, E_b/N_o for MFSK approaches the Shannon bound of -1.59 db-Hz and bandwidth efficiency approaches zero. If we now look at 2PSK, $E_b/N_o = 9.6$ dB-Hz for BER = 10^{-5}. This is 11.2 dB short of the Shannon bound.

For QAM, when QAM and PSK have the same value of M ($M \geq 8$), QAM modulation allows operation at lower values of E_b/N_o for the same bandwidth efficiency, E_b/N_o in dB-Hz.

Also superimposed on the curve is a condition for coding on E_b/N_o and bandwidth efficiency when the channel is *power-limited*. Error control can be used to expand bandwidth to allow operation at lower values of E_b/N_o. Therefore, to realize the desired P_B, we must move to the right and downward in Figure 2–1 (resulting in lower R/W ($W = B$). For example, for convolutional encoding and Viterbi decoding for a rate of 1/2, and using QPSK, the bandwidth efficiency drops to "1" from the value of "2" shown. Conversely, the code rate of 1/2 requires twice the bandwidth of the uncoded system design.

Table 2–1 It is obvious that bandwidth efficiency is improved as the number of modulation states increases. Extra energy is required, but in bandwidth-limited systems this may be worth the exchange.

Modulation	E_b/N_o	R_b/B
8PSK	14.0	3
8QAM	10.6	3
16PSK	18.3	4
16QAM	14.5	4
64PSK	26.5	6
64QAM	18.8	6

When a channel is *bandwidth-limited*, it is desirable to utilize modulation techniques that allow the transmission of more bps in a given bandwidth (i.e., higher R/W). Therefore, using a higher alphabet size with M-PSK is desirable; or better still is the use of M-QAM. As we have indicated, M-QAM allows greater generation at lower values or average E_b/N_o than M-PSK when $M = 8$.

It is interesting to note from Figure 2–1, but not explicitly indicated therein, that as the modulation order is increased, and the information rate is left *unchanged*, the power efficiency will increase. First note that the points shown plotted for M-ary PSK and M-ary QAM are values only for discrete BER values. Actually, these points fall on a series of trajectories in the capacity plane for different values of E_b/N_o. A sampling of some of the points in Figure 2–1 are depicted in Figure 2–2. The point we are trying to make here is that we can maintain the *same* channel capacity of x bits / s-Hz by going to higher-order modulation methods and by reducing the energy requirement. This is achieved without coding the modulation (for example, for 8-PSK and 16-PSK). We can achieve the same bandwidth efficiency of 3 bits / s-Hz for 16-PSK as 8-PSK, but at an E_b/N_o of roughly less than 6 dB. If 16-QAM is used, E_b/N_o is about 8 dB less. This is represented on the curve.

In coding, we can maintain the *same* BER by increasing the power efficiency, but we experience a decrease in bandwidth efficiency. As a counter to increasing the bandwidth, we can increase the number of *coded* symbols instead of the bandwidth. Therefore, the channel alphabet is increased, but the information rate (as opposed to moving down the slope of Figure 2–2) is held fixed. The incorporation means less power is required to maintain the BER.

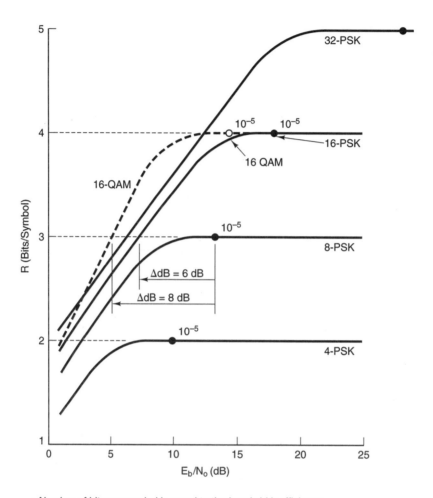

- Number of bits per symbol is equal to the bandwidth efficiency.
- When the channel alphabet (e.g., 4PSK → 32PSK) increases, and the information rate is left unchanged, the power efficiency will increase.

Figure 2–2 Capacity of several higher-order modulation waveforms. Copyright © 1984 by the Institute of Electrical and Electronics Engineers, Inc. Reprinted with permission.

Some of the comments made above in reference to Figure 2–1 can be summarized in Figure 2–3 [4]. Various tradeoffs can be made by the migration of the points in the plane to the left of the Shannon curve. These excursions occur by the manipulation of E_b / N_o, BER, or B (that is, R / B).

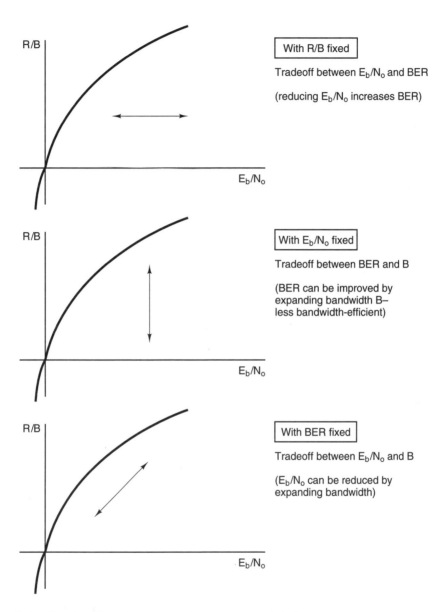

Figure 2–3 Trade-off performances in various modulations.

2.3 Conclusions

In pursuit of the Shannon bound, Shannon postulated that the key to efficient transmission of data was not found only in the S/N, but in how the data was encoded.

The demand for improved communications has led to the concept of signal coding. As the information in this chapter indicates, we have yet to reach the Shannon bound. For example, with reference to 8-PSK on Figure 2–1, the E_b / N_o required to realize 10^{-5} is 13 dB-Hz — a shortfall of 13 + 1.59 = 14.59 db-Hz from the Shannon bound. There is some improvement by using M-ary QAM, but it is not significant. The solution in reaching the Shannon limit is to use coding. The whole area of coding has been and is continuously being explored in an attempt to reach the Shannon bound. Even though some progress has been made, much still remains to be done. To reach the channel capacity limit, highly complex coding schemes are required. It should also be remembered that a good theoretical code does not necessarily work in practice.

In Chapter 3 we will show that higher order modulation methods via M-ary signaling (multi-level signaling) can convey additional digital information. We can improve the bandwidth efficiency (bits/s-Hz) by increasing the number of symbols states (even into the hundreds or more). However, the practical limit is determined by the ability of the receiver to resolve the states in a hostile environment (noise, interference, etc.). We will discuss the pros and cons of using higher order modulation states.

Table 2–2 depicts the error probabilities and E_b/N_o required for several modulations.

Table 2–2 E_b/N_o^* (dB) of Various Bit Error Rate Probabilities

p(e)	Binary PSK (BW = R)	Quaternary PSK (BW = R/2)	8-ary PSK (BW = R/3)	16-ary PSK (BW = /4)
10^{-1}	–0.9	–0.9	4.0	8.50
10^{-2}	4.3	4.3	8.0	12.70
10^{-3}	6.80	6.80	10.20	15.00
10^{-4}	8.40	8.40	11.80	16.40
10^{-5}	9.60	9.60	13.00	17.60
10^{-6}	10.50	10.50	13.90	17.60
10^{-7}	11.30	11.30	14.70	19.30
10^{-8}	12.00	12.00	15.50	19.80

* $(E_b / N_o) = (S / N)(BW / R)$
E_b / N_o dB = S / N dB + 10 log (B / R) dB
S / N dB = E_b / N_o dB + 10 log (R / B)
$R = 1 / T$, $N = N_o B$ = AWGN, $E_b = ST$ joules = watt seconds

2.4 References

[1] C.E. Shannon. "Communications in the Presence of Noise," *Proc. IRE*, January 1949.

[2] C.E. Shannon and W. Weaver, "The Mathematical Theory of Communication," Univ. of Illinois Press, Urbana, IL, 1962.

[3] B. Sklar, "Defining, Designing and Evaluation Digital Communication Systems," *IEEE Comm. Magazine*, November 1993.

[4] E. Biglieri, "High Level Modulation and Coding for Nonlinear Satellite Channels, *IEEE Trans. Communications*, May 1984.

[5] M.P. Ristenbatt, "Alternatives in Digital Communications," *Proc. IEEE*, June 1973.

[6] G. Ungerboeck, "Channel Coding and Multilevel/Phase Signals," *IEEE Trans on Information theory*, January 1982.

[7] R.S. Burington, *Handbook of Mathematical Tables and Formulas*, Handbook Pub. Co., Sandusky, OH, 1958.

CHAPTER 3

Higher Order Modulation Methods

3.1 Introduction

In recent years, we have been trying to obtain more and more telecommunications services out of a limited amount of spectrum. As a result, channelization in the various systems is becoming bandwidth-limited. In an attempt to improve this paucity, modulation methods that offer greater bandwidth efficiencies are being used. These higher order modulation methods have found applications in terrestrial microwave radio, and more recently, in satellite and wireless communications. The intelligent application of these digital modulation techniques provide the means of coping with this spectral scarcity.

The improvement in bandwidth efficiency, that is, the transmission of more bits per second in a given bandwidth, or greater bit rate per bandwidth, (R / B), is presented via higher order modulation schemes, including M-ary PSK and M-ary QAM. Both realize good bandwidth efficiencies. Where M-ary QAM offers comparable theoretical bandwidth efficiencies as M-ary PSK for small values of M ($M \leq 8$), QAM offers lower values of E_b / N_o for the same BER. This is demonstrated by Figure 3–1. For example, for BER = 10^{-5}, a 16-level PSK signal will require an E_b / N_o = 17.2 dB. For the same error rate and 16-level QAM, E_b / N_o = 13.4 dB for the same performance. There is therefore an improvement of 3.8 dB for 16-QAM operation. This is due in large measure to the larger signal distance achieved between adjacent signal states within the constellation in the signal state space diagram. More will be said on this later.

As indicated in both displays, higher order modulation methods require more energy. Therefore, we are trading bandwidth efficiency for an increased RF power

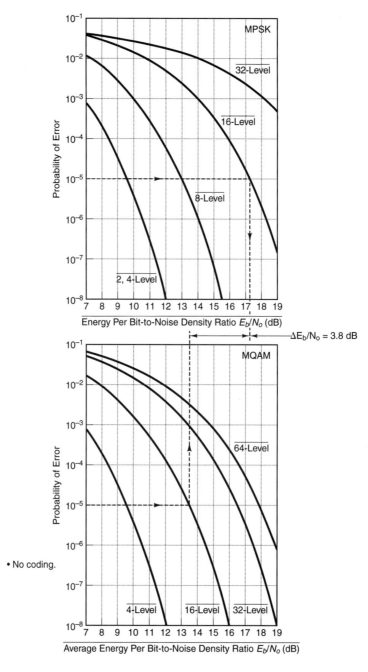

Figure 3–1 BER vs. E_b/N_o comparison of M-ary PSK and M-ary QAM modulations.

requirement. That is, we are dealing with a bandwidth-limited as opposed to a power-limited system.

It should also be noted that the data indicated in Figure 3–1 is for linear operation of both QAM and PSK. QAM signals in filtered and nonlinear channels can give worse performance than PSK signals, even though QAM performs better in a linear channel. However, this does not mean that the utility of M-QAM is limited because linearizers to correct this problem are available and are used. For example, using adaptive transversal equalizers, or backing-off power amplifiers into the linear region of their operation, both alleviate this limitation.

Some bandwidth efficiencies realized from M-ary modulation are indicated in Table 3–1. However, there are caveats, and theoretical numbers will not be achieved in practice. Efficiencies of 10 – 20% less than those shown appear to be feasible and are usually realized.

Figure 3–2 shows the power efficiency of M-ary QAM ($M \geq 8$) over M-ary PSK for the same bandwidth efficiency, R/B. These are practical numbers. Also shown is the Shannon bound, or the absolute maximum capacity which would be extremely difficult to realize in practice, if ever. Notice that for BER = 10^{-4}, the shortfall is about 7 dB. It is of interest to note also that modulation that depends only on phase information (M-PSK) is less efficient than those methods that use both phase and amplitude (M-QAM). This appears to be another example that modulation alone will not bring you much closer to the Shannon bound, but needs additional help from *coding*.

When we have a power-limited condition in the channel, error control coding can be used to expand bandwidth to allow operation at lower values of E_b/N_o (for the same BER); however, a concomitant reduction in bandwidth efficiency (R/B) is present. More modern modulation techniques allow us to realize a coding gain *without* the increase in bandwidth, that is, using the elegant Trellis Coded Modulation (TCM). This subject will be presented in another chapter.

As alluded to above, the improvement in M-QAM over M-PSK for higher order modulation methods is easily demonstrated by using signal state space diagrams. For example, Figure 3–3 depicts the signal state space diagrams for 16-PSK and 16-QAM.

In the constellation for 16-PSK (an alphabet of $M = 16$ symbols), the number of transmission bits per symbol is four, $n = \log_2 M = \log_{10} M / \log_{10} 2$. In general, for n bits/symbol, two states are required. Therefore, for 16-PSK and four bits per symbol, $2^4 = 16$ states are required.

In M-ary QAM, each state point is represented by a symbol made up of several bits. For n bits per symbol, the number of states required to represent the information is $M = 2^n$. Therefore, for 16-QAM, the number of states is $M = 2^4 = 16$, where $n = 4$.

Notice that as the order of the modulation increases in M-PSK, the points become increasingly closer, and the signal points become increasingly vulnerable to perturba-

Table 3–1 Comparison of Bandwidth Efficiencies and Other Information for Different Modulations.

Modulation M-ary signaling*	Number of States (M logic Levels)	Theoretical Bandwidth Efficiency** bits/s – Hz	Bits Sent each Time XMTR is Keyed (each Symbol)	Required Bandwidth Hz
2-PSK	2	1	1	bit rate
4-PSK (QPSK)	4	2	2	1/2 bit rate
8-PSK	8	3	3	1/3 bit rate
16-PSK	16	4	4	1/4 bit rate
32-PSK	32	5	5	1/5 bit rate
16-QAM	16	4	4	1/4 bit rate
64-QAM	64	6	6	1/6 bit rate
256-QAM	256	8	8	1/8 bit rate
512-QAM	512	9	9	1/9 bit rate
1024-QAM	1024	10	10	1/10 bit rate

* M-ary signaling is any signal scheme where the number of possible signals sent during any given signaling interval is M. Binary signaling is a special case with $M = 2$. MPSK uses M phases of a sinusoidal carrier. Note for QAM, signaling includes both amplitude and phase.

** Efficiency will be affected by the channel Nyquist cosine rolloff filter. The rolloff factor is given as "α" and lies in the range $0 \leq \alpha \leq 1$. Practical values of alpha are 0.2 – 0.5. The theoretical efficiency is divided by $1 + \alpha$. "Brick wall" filters have an alpha equal to zero. Transmit and receive filters act in unison and collectively give an effective "α".

tions in the channel. That is, an amplitude or phase disturbance may be interpreted, by the detection process in the receiver, as the adjacent symbol.

For M-ary PSK modulation, the distance between the states, d, can be found to be equal to

$$d = 2A\sin(\pi / M) \qquad (3.1)$$

where M is the number of states, and A is the peak amplitude of the signal,[1] for M-PSK.

---

1. Some texts on the subject use the term "signal phasor length" for A.

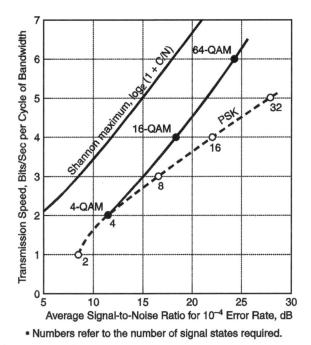

Figure 3–2 Spectral efficiencies for M-ary PSK and M-ary QAM for BER = 10^{-4}, showing the shortfall from the Shannon bound.

A is constant since the points lie on a circle about the orgin. For simplicity, A is usually normalized to equal unity.

For 16-PSK, from Equation (3.1), we calculate the distance $d = 0.393A$. This has frequently been referred to as the *Euclidean* distance.

In M-QAM, each state point is represented by a symbol made up of several bits. By using an alphabet of M symbols, the transmission of $n = \log_e M$ bits during each symbol period is feasible. For 16-QAM, we have $n = 4$, or four bits per symbol.

For 16-QAM, the Euclidean distance is shown to be equal to

$$d = (2/3)A\sin 45° = 0.47A \tag{3.2}$$

where A is the signal phasor length.

Comparing the Euclidean distance of 16-PSK and 16-QAM shows that it is greater for 16-QAM. The implication is that it will be less vulnerable to channel impairments like amplitude and phase distortions. Also, this increased distance is reflected in a better BER for 16-QAM over 16-PSK for the same degree of interference. This has been shown previously in the "waterfall" curves (Figure 3–1) where $M = 16$. 16-QAM

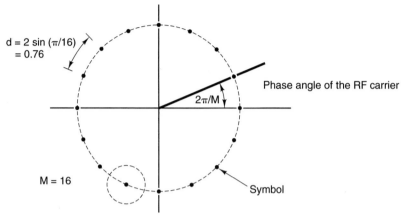

- Constant envelope signal.
- All the information must be encoded in the phase.

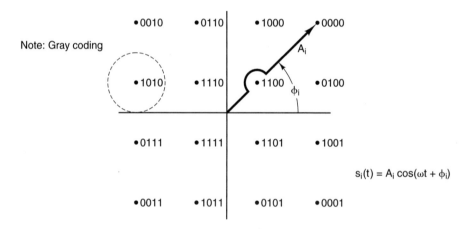

- Non-constant envelope.
- Information is contained in both the amplitude and phase of the carrier.
- The improved robustness of 16-QAM over 16-PSK is due to the greater Euclidean distance.

Figure 3–3 Signal statespace diagrams (constellations) for 16-PSK and 16-QAM.

achieves the same BER as 16-PSK with 4 dB less E_b/N_o. The error performance of any digital modulation system is fundamentally related to the distance between points in the signal constellation.

It is interesting to note that if the channel alphabet is *maintained* and the *information rate* (R/B) is held constant (see Figure 3–2), the increased Euclidean distance will increase the power efficiency. It may be of further interest to refer back to Figure 2–2,

Signal State Space Diagrams

which shows a family of M-PSK curves. If the size of the alphabet is *increased*, but the information rate is left *unchanged*, there is an increase in power efficiency. Therefore, in going from 8-PSK to 16-PSK, for a $R/B = 3$ b/s-Hz, an improvement of 6 dB can be realized (as indicated), *but* at an increase in BER — a subtle point.

3.2 Signal State Space Diagrams

The idealized signal state space diagrams, where $N = \phi$ for higher order modulation types are represented in Figure 3–4. For M-ary PSK, the signal is of a constant amplitude and the information is contained in the phase. In M-ary QAM, the information is contained in both the amplitude and phase. Therefore, the constellation for M-ary QAM usually results as a rectangular form of plot. However, two additional forms are possible and are shown for 16-ary QAM. Each signal state symbol differs from its adjacent state symbol by one bit. This is referred to as Gray encoding. Note, for example, the points in the 16-QAM constellation.

Each symbol in an M-ary alphabet is related to a unique sequence of n bits transmitted in T_s seconds.

$$M = 2^n \text{ or } n = \log_{10}M = \log_{10}M / \log_{10}2 \tag{3.3}$$

Since one of M symbols is transmitted during each symbol duration, T_s, the *data rate*, R in b/s, can be expressed as

$$R = n / T_s = \log_2 M / T_s \tag{3.4}$$

Data bit duration, T, is the reciprocal of the data rate as related to symbol rate, R, as follows:

$$T = 1 / R = T_s / n = 1 / nR_s \tag{3.5}$$

The *symbol rate*, R_s, in terms of the *data bit rate*, R, is therefore

$$R_s = R / \log_2 M \tag{3.6}$$

In the 16-QAM constellation, there is approximately a 10 dB difference in power between the minimum and maximum carrier (vector) levels. The implication of different amplitudes is that the channel must be linear to prevent one signal state from being interpreted as another signal state. Any vacillation in phase will also move a signal point into the decision domain of another point.

Actually, *point* displays in signal space are a practical fiction since each point will be fuzzy (scatter plot) because of noise and impairments in the system. The display of two noisy signal points in state space, which are in proximity to each other, is shown in Figure 3–5. Bona-fide signals are located at the center of the "piles". If the noise

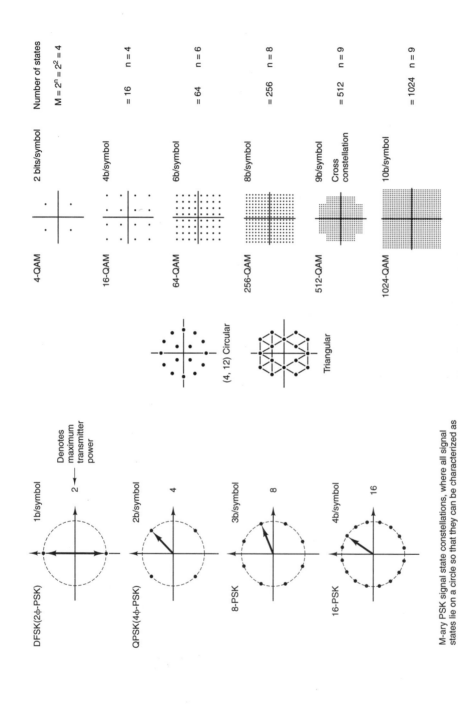

Figure 3–4 Constellations for several modulation states for M-PSK and M-QAM.

(roughly the rms value) exceeds one-half the distance to the adjacent point, it is possible that the adjacent state will register at the detector. Therefore, an error occurs. One can spread the symbols apart by increasing the signal power if additional power is available, or alternately improve the noise level of the channel. We should note that the Gaussian display indicates both amplitude error *and* phase error. Clearly, you may encounter a situation where one or the other may dominate.

Different impairments reveal different displays and information as to the nature of the perturbation. Some typical interference patterns in signal space are shown in Figure 3–6. Figure 3–6(a) shows a constellation for a benign channel. The signals are fixed in amplitude and phase. In Figure 3–6(b), noise (with Gaussian statistics) is additive to the channel, and perturbs both the phase and amplitude. This was shown previously in Figure 3–5.

In Figure 3–6(c), phase smearing occurs, and there is a vacillation about the desired phase value, resulting in distortion of the ideal 90° between I and Q. In Figure 3–6(d), there is a radial movement of the signal levels, indicating amplitude changes, possibly due to channel non-linearities and impairments.

In general, for all cases indicated in Figure 3–6, the closer the signal states are to each other (Euclidean distances), the more vulnerable the system impairments, the larger the BER, and consequently the larger the probability of the bit error rate (P_e). The amount of interference which can be tolerated is a function of the BER required. Clearly, very small BERs (like 10^{-9}) may be difficult to achieve, if not impossible to obtain with higher order modulation types such as 256-QAM and 512-QAM. This is caused by the extreme proximity of the signal states.

3.3 Performance Representations

The advent of bandwidth-efficient modulation techniques has sprouted several measuring procedures to determine performance in the presence of channel impairments. Four basic methods are normally used:

1. *Spectrum analysis* displays spectral shape, including spectral sidelobes and bandwidth. However, this method may not reveal certain types of impairments which may affect the signal.
2. An *eye diagram* is displayed on an oscilloscope. It is referred to as an eye diagram because it resembles the eye for binary digital signals. It enables amplitude and phase error measurement.

    Eye diagrams are generated by driving the vertical input of an oscilloscope by the binary digital signal; the horizontal input synchronizes the baseline to the period of the bit. This setup is shown in Figure 3–7(a). Also shown are two digital drive waveforms. For example, the ideal binary waveform BPSK

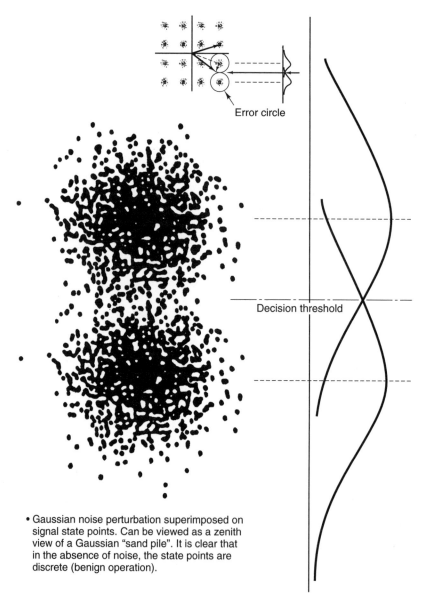

- Gaussian noise perturbation superimposed on signal state points. Can be viewed as a zenith view of a Gaussian "sand pile". It is clear that in the absence of noise, the state points are discrete (benign operation).

Figure 3–5 Noise channel impairments depicted in signal space.

with perfectly square bits (Figure 3–7(b)) is represented on the oscilloscope as two horizontal lines with the top line representing bit "1" and the bottom line bit "0". For a practical bandwidth-impaired signal, shown in Figure 3–7(c), the

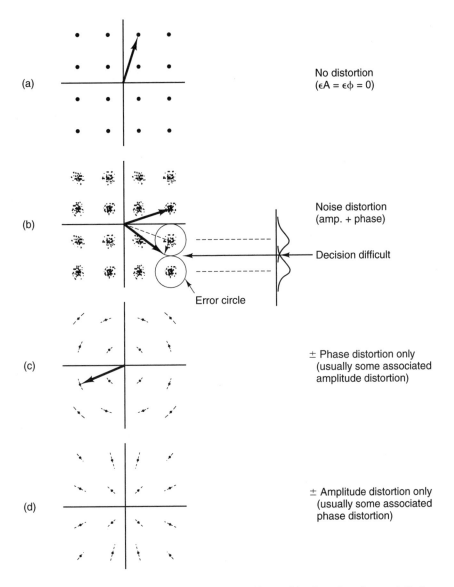

Figure 3–6 Various channel impairments manifested in the signal constellations.

pulses are rounded off and the eye becomes clearly visible. As will be demonstrated later, the opening of the eye and its points of zero crossover indicate the measure of performance. For now, in simple terms, the invariance of the crossover (no time jitter), with 100% eye opening, indicates distortionless transmission.

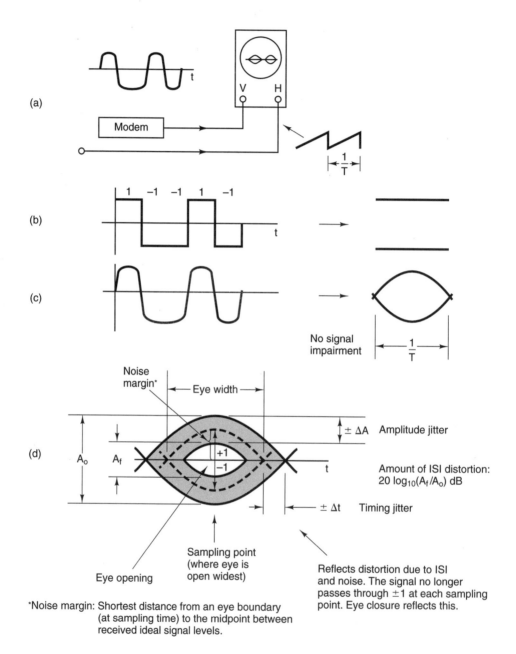

Figure 3–7 Eye diagram generation and the performance information it reveals.

A display of multilevel M-ary QAM modulations is depicted in Figure 3–8; both their constellations and their associated eye diagrams are shown. Their associated theoretical bandwidth efficiencies are indicated on the left. At the present time, 64-QAM is in the operational inventory. As we indicated previously, the higher the modulation order, the more difficult the design becomes because of the proximity of the signal points in the constellation.

The points on the I-Q plane are shown as discrete points indicating that there are no channel impairments. Actually, the eye diagrams shown do not quite reflect the constellation since the eyes are experimental results and manifest some impairments. We would therefore have smearing (and possible rotation) of the signal dots.

Referring to Figure 3–7(d), the practical eye pattern for a binary waveform is not a single trace on the oscilloscope. Perturbations in the channel cause the eye to "close" and narrow in the horizontal direction. These phenomena are caused by amplitude and phasing (or timing) jitter errors, respectively. Note that when the eye is partially closed, the decision as to whether a "1" or a "0" was received becomes more difficult. Similarly, timing jitter moves the apex of the eye to either the left or right at the sampling point, and as a result, the amplitude at the decision threshold varies.

Since eye diagrams and signal constellations reveal system performance, it is of interest to correlate the constellation scatter display to the eye diagram. Figure 3–9 suggests this correlation. The scatter is caused by errors whose diameter relates to eye closure. Clearly when the closure reaches the decision axis, system performance has degraded to the point where multiple bit errors are manifested.

3. *Signal state space diagrams* are also referred to as signal constellations. This has been discussed previously and will not be amplified here. However, there is one point which was not mentioned above which may be of interest. That is, sometimes the signal points display may be asymmetrical. This is indicative of an imbalance ($\Delta\phi \neq 90°$) in the phase and quadrature channels.

4. The *vector diagram* displays the I and Q components on an oscilloscope, which reveals the *dynamics* of the symbol transitions. Alternate paths in the transition from one symbol to another reveals the modulation quality. A void of errors would show direct transitions to different signal points. An example is shown for $\pi/4$-QPSK in Figure 3–10. The top-most display shows direct transitions for a perfect design. The bottom-most display shows the alternate paths for a non-ideal condition, which is found in operational-implemented systems.

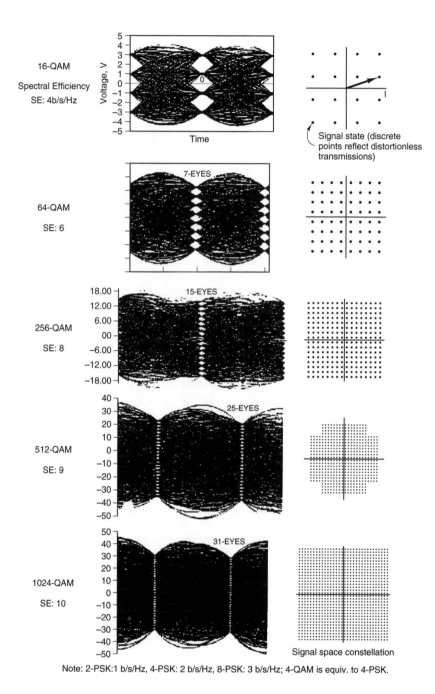

Figure 3–8 Eye diagram and signal state space constellations for M-ary QAM modulations. Copyright © 1985 by the Institute of Electrical and Electronics Engineers, Inc. Reprinted with permission.

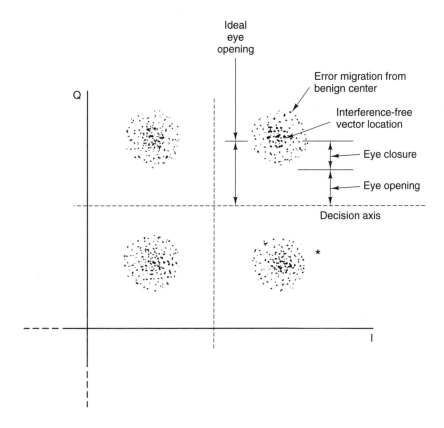

*Can be looked at as a two-dimensional Gaussian density function as viewed from its zenith.

Figure 3–9 Eye diagram "translatable" to signal space plane.

3.4 Conclusions

M-ary PSK modulation with bandwidth efficiencies on the order of 2 b/s-Hz is still a popular modulation technique used in existing communications systems. However, the scarity of spectrum has made it necessary to pursue more power and bandwidth-efficient (bits/s-Hz) modulation schemes. A format which is more bandwidth-efficient is M-ary QAM. 16-QAM and 64-QAM are also presently being used in terrestrial wireless systems. A listing of some of these systems' efficiencies can be found in Table 3–1.

30   Ch. 3 • Higher Order Modulation Methods

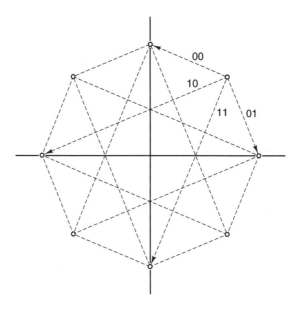

(a) Constellation for π/4-QPSK, for ideal operation.

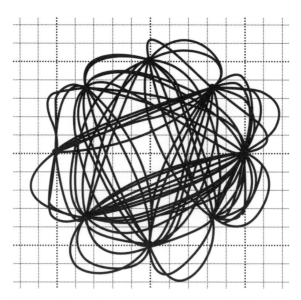

(b) Vector diagram for π/4-QPSK, depicting symbol dynamics.

Figure 3–10   Alternate representation of digital signal.

However, there are a few caveats to consider when selecting a modulation scheme. Generally, more power, i.e., E_b / N_o, is required to go to the higher order modulation formats (see Figure 3–1). In addition, QAM is less robust than BPSK and QPSK and more vulnerable to channel impairments because it is not a constant envelope signal. Therefore, linear channels or compensation for non-linear channels is required to realize the performance desired. There are, however, palliatives which can be incorporated to maintain performance. These include power amplifier backoff into the linear region (at a sacrifice of output power). Equalizers/linearizers are also used extensively.

We can receive alleviation from this shortfall in power by the use of Forward Error Correction (FEC). But here too we must be careful because there is an erosion in bandwidth efficiency. Here, Trellis Coded Modulation (TCM) may come to the rescue by offering coding gain without an increase in bandwidth. This concept will be discussed in Chapter 6.

Research in the laboratory is presently looking into still higher order modulation formats; in particular, 256-QAM, 512-QAM, and even 1024-QAM. In fact, 1024-QAM is used in 56 kbps modems. However, channel purity is increasingly difficult to realize to cope with extremely small Euclidean distances. Only time will tell as to whether we will see these in practical systems.

In Chapter 4 we will discuss some of the modern digital modulation techniques used in wireless communications. Some manifest efficiency in terms of spectrum and others exhibit efficiency in terms of power requirements. However, either way, they all have application depending on the design rules applied.

3.5 Glossary of Terms

- E_b = energy per bit
- E_s = energy per symbol
- $E_s = E_b \log M = E_b(\log M) / \log 2$
- M = size of alphabet (or number of states), e.g., 4-PSK = 4 output phases, $M = 4$
- $n = \log M$, number of bits per symbol e.g., $M = 4$, $n = \log 4 / \log 2 = (0.6/0.3) = 2$
- Symbol rate bit rate, $R_s = R_b / \log M$
- If $M = 2$ signal pulses can be sent every switching time, T, then the data rate is n / T bits per second. If there are n bits/symbol, there must be M different symbols to uniquely convey all bit sequences.
- Symbols: member of the M-ary alphabet that is transmitted during each symbol duration T_s. One of M symbols is transmitted during each symbol duration. One or more bits can be sent as a symbol. Symbol may be sent as a digital level.

3.6 References

[1] B. Pattan, "Spectrally Efficient Higher-Order Modulation Architectures for Terrestrial and Satellite Applications," *FCC/OET Technical Report*, June 10, 1991.
[2] J.C. Bellamy, *Digital Telephony*, Wiley and Sons, New York, 1982.
[3] F. Xiong, "Modern Techniques in Satellite Communications," *IEEE Comm. Magazine*, August 1994.
[4] B. Buxton, "Measurement Methods Analyze Digital Modulation Signals," *Microwaves and RF*, July 1995.
[5] D. Rockwell, "AT&T Introduces 64-QAM Digital Microwave Radio," *Microwaves and RF*, August 1984.
[6] B. Pattan, "Graduate Courses in Communication Theory and Short Courses at GWU on Modulation Theory," 1993, 1994.
[7] P. Mathiopoulos, et al., "Performance Evaluation of a 512-QAM System in Distorted Channels," *IEEE Proceedings*, Part F, No. 2, April 1986.

CHAPTER 4

Dynamics of Linear and Continuous Phase Modulation Methods in Digital Communications

4.1 Introduction

During the last decade the burgeoning growth of terrestrial wireless communications and satellite services has prompted the pursuit of new modulation techniques that provide both spectral efficiency, because of the paucity of spectrum and communication efficiency, because of power limitations. These investigations worldwide have proceeded in the direction of digital communications. New digital concepts which are being entertained include continuous phase modulation (CPM), which has desirable attributes in non-linear channels for both terrestrial mobile and satellites. CPM is essentially frequency modulation (FM) in which the baseband modulating signal is a digital signal. We will address some of these concepts in this chapter.

The increase in telecommunications services being provided in the U.S. and globally requires a re-examination of the available frequency spectrum required to provide these services. Clearly, there is a paucity of spectrum, and methods have been devised to accommodate the burgeoning services.

Several artifices have been exploited to accomplish this end, including, on macroscopic scale, frequency reuse where the effective bandwidth available is over and above the allocated frequency bandwidth. One method used to accomplish this is space division multiplexing, that is, assigning the same spectrum to different geographical locations that are sufficiently isolated to prevent mutual interference. Another more recent concept is to assign frequencies on a time-shared basis. Clearly, this requires synchronization among the various users. One popular scheme is to partition the spectrum, which allows users exclusive use (segmentation). However, this limits the amount of spectrum assigned to individual users. Another scheme is the assignment of parts of the electromagnetic spectrum which lie fallow. This usually means using the higher bands in the spectrum.

On the microscopic scale, we can revert to the design of the signal itself so that more channels can be packed into a limited spectrum. The trend is therefore to digital modulation schemes, which can be more spectrum-efficient and power-efficient than analog modulation.

In digital signal design, it is desirable to have the following attributes:

- <u>High spectral efficiency</u>. Various investigators have defined this concept in different ways. Here we define it as bits/s-Hz, that is, the maximum number of bits which can be packed into an allocated bandwidth. If R is the data rate in bits/second and B is the bandwidth occupied by the modulated RF signal, the spectral efficiency is expressed as

$$\eta_B = R / B \quad \text{b/s-Hz}$$

  Clearly, the greater the spectral efficiency, the greater the capacity of the system, with power limits notwithstanding. That is, higher order modulation systems (e.g., M-ary PSK, M-ary QAM) generally require higher E_b / N_o to realize the desired BER. Where E_b is energy per information bit and N_o is the system noise power density.

- <u>Higher communication efficiency</u>. This is the lowest possible E_b / N_o to realize the desired BER, not including FEC coding, which consumes bandwidth, at least in the classical sense, unless one uses Trellis Coded Modulation (TCM), where coding gain is achieved without a sacrifice of bandwidth.

- <u>Low adjacent channel interference (ACI)</u>. This is invoked by the distribution of the signal spectrum (main lobe + sidelobes) and its representation in the time domain. Too narrow a signal spectrum may cause inter-symbol interference since it reflects on a wider pulse in the time domain.

- <u>Use of a signal design which is robust in a non-linear channel</u>. This will be addressed in subsequent sections.

We will confine our analysis in this chapter to signal designs which provide good spectral efficiency, low spectral sidelobes, and power efficiency.

4.2 Linear Modulation

There are basically two classes of signal modulations used in digital communications. The first is linear modulation, which consists of modulations of the following types:

Linear Modulation

- Binary PSK.
- QPSK.
- Offset QPSK.
- $\pi/4$-QPSK.
- M-ary PSK.
- M-ary QAM.

The other is constant envelope modulation (nonlinear modulation). These types include

- FSK (Frequency Shift Keying).
- MSK (Minimum Shift Keying.
- GMSK (Gaussian Minimum Shift Keying).
- Tamed FM (TFM).
- Generalized Tamed FM (GTFM).

Frequency shift keying exhibits a constant envelope, where the transmitter selects a range of frequencies or tones during a particular symbol interval. To maintain continuous phasing requires that a single VCO be keyed. Switching oscillators to produce the tones would not produce continuous phasing. The other FSK-like modulation types (above) are variations of digital FM, having desirable properties which are both spectral- and power-efficient.

Linear modulation schemes generally have higher spectral efficiencies than constant envelope modulation types. Typically, $n = 1$ b/s-Hz for BPSK, $n = 2$ b/s-Hz for QPSK, and higher for higher order modulations (M-ary PSK, M-ary QAM). QPSK are most commonly used in satellite communications, with the higher order modulation types used in terrestrial systems where there is adequate power available. In the earlier years of satellite technology, the systems were power-limited and spectrum was more readily available. However, with higher power coming into existence, and additional services provided via proliferation, spectrum is scarce and higher spectral efficiencies are now in high demand.

Linear modulation types have a shortcoming which requires that any kind of limiting or non-linear amplification be prevented. Linear Class-A amplifiers are therefore required. To reduce the out-of-band emissions caused by the spectral sidelobes, it is required that a signal be band-pass filtered. However, when shaping is used to confine the spectrum, the signal loses its constant envelope property. To restore the constant envelope, it is necessary to limit the signal. The amplitude-amplitude (AM/AM) characteristic of the non-linearity tends to restore the signal to constant envelope. This operation introduces the anomalous behavior by restoring the filtered spectral sidelobes and therefore spreads the transmitted signal. Clearly, this is an unacceptable situation. The underlying case of this anomaly is the impulsive phase changes during the bit transitions. To prevent this regrowth, the signals must be amplified by linear amplifiers. The more efficient Class-C amplifiers cannot be used because of their non-linear operation. Typical efficiency of Class-C amplifiers is 70%. That is, 70% of the applied DC power is converted to RF. On the other hand, Class-A amplifiers (required in linear modulation) have efficiencies on the order of 40%.

The bit phase transitions for various linear modulations are indicated on the signal-space diagram illustrated in Figure 4–1. The symbol transitions for BPSK are indicated in Figure 4–1(a). It is noted that for every transition between 1 and 0, there is a ±180° phase change. During these transitions, the signal envelope goes to zero. For QPSK (Figure 4–1(b)), the constellation of symbols is deployed in the four quadrants of the I-Q plane. Any transitions that cross through the origin will produce envelopes which drop to zero. Any bit changes from one quadrant to the adjacent quadrant will produce ±90° phase changes. The divots in the envelope will not be as deep, but they will drop to about 70% of the maximum value, or a loss of 3 dB.

The offset QPSK (OQPSK) in Figure 4–1(c) also demonstrates phase transitions of 0° and ±90° maximum with no transitions through zero. This is an improvement of the phase transitions over QPSK.

The modulation π/4-QPSK in Figure 4–1(d) also undergoes fluctuations, but not as severe as QPSK. π/4-QPSK is a form of QPSK modulation in which the QPSK signal constellation is shifted by 45° for each symbol interval, T. The phase transitions from one symbol to the next are limited to ±π/4 and ±135° — not quite as severe as BPSK and QPSK. The 135° shift is equivalent to an attenuation of 20log(100/38) = 8.4 dB.

The time display of these envelope fluctuations are shown in Figures 4–2 and 4–3 for BPSK, QPSK, and OQPSK. In Figure 4–2, note that for every bit change, the envelope goes to zero. In Figure 4–3(a), QPSK shows both 180° phase dips and 90° phase dips. Figure 4–3(b) shows the time display for OQPSK with only 90° dips in the envelope. As we would expect, the smaller the phase dips, the less susceptible the waveform to spectrum regrowth after non-linear amplification.

The spectral regrowth of the spectral sidelobes of digital waveforms is vividly demonstrated in Figure 4–4. The top-most figure is the theoretical power spectral density for BPSK and OQPSK (or QPSK). The bottom-most figure shows output spectra after band-pass filtering and limiting, or non-linear amplification. It is clear that BPSK has undergone aspectral regrowth. For the OQPSK signal, even though the original spectrum (prior to filtering) displays significant sidelobes, filtering and limiting do not restore the spectral sidelobes. This further confirms the requirement that BPSK must be amplified by linear amplifiers to prevent the regeneration of the sidelobes.

Quadrature modulation generation of QPSK and OQPSK is indicated in Figure 4–5. A binary bitstream with $R = 1/T$ is encoded to a rectangular NRZ signal. The bitstream is converted into two bitstreams, $a_I(t)$ and $a_Q(t)$, each with a rate of $1/2T$. $a_I(t)$

Linear Modulation

(a)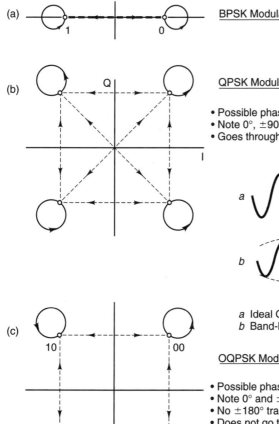

BPSK Modulation Constellation

(b)

QPSK Modulation Constellation

- Possible phase states of the QPSK-modulated carrier.
- Note 0°, ±90°, and ±180° are possible.
- Goes through origin (bad).

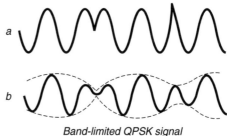

Band-limited QPSK signal

a Ideal QPSK
b Band-limited QPSK

(c)

OQPSK Modulation Constellation

- Possible phase states of the OQPSK-modulated carrier.
- Note 0° and ±90° phase transitions are possible.
- No ±180° transitions.
- Does not go through origin.

(d)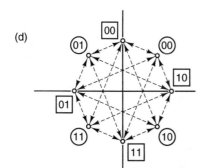

π/4-QPSK Modulation Constellation

- Possible phase states of the π/4-QPSK-modulated carrier.
- Maximum phase change is ±135°.
- Eight phases are possible, but only four are active at any one time.
- Uses Gray coding for modulation.
- Does not cross origin.
- Appears to be an eight-phase pattern, but it is not.
  Signal elements of the modulated signal are selected in turn from two QPSK constellations, which are shifted by π/4 with respect to each other.

Figure 4–1 Signal state space diagrams for linear modulations.

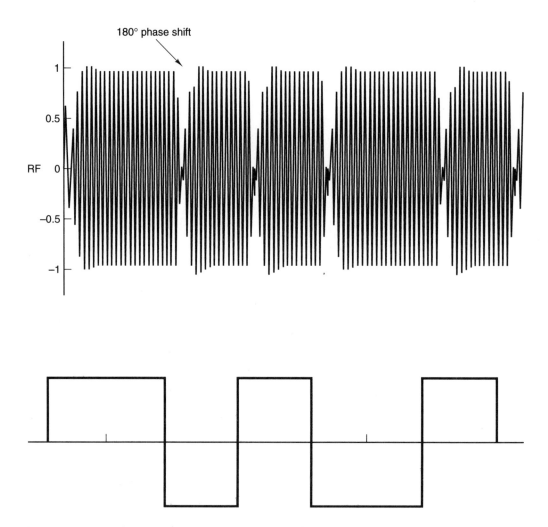

Figure 4–2 BPSK modulation showing the 180° phase transitions.

and $a_Q(t)$ are converted into parallel bitstreams each with a rate of $1/2T$. The $a_I(t)$ stream modulates with $\cos\omega 2f_c t$ in the I channel and $a_Q(t)$ modulates with $\sin 2\omega f_c t$ in the Q channel. These are added by the summer to give

$$s_{QPSK} = a_I(t)\cos(2\omega f_c t) + a_Q(t)\sin(2\omega f_c t) \qquad (4.1)$$

Depending on the values of $a_I(t)$ and $a_Q(t)$ (note these are simply ± multipliers), the phase can assume values of 0°, 90°, 180°, or 270°. For QPSK, as indicated previously, the phase values may assume values of 0°, ±90°, or ±180°. For example, a change in

Linear Modulation

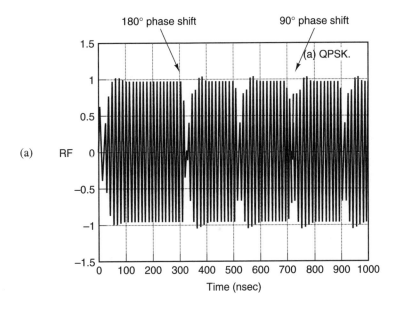

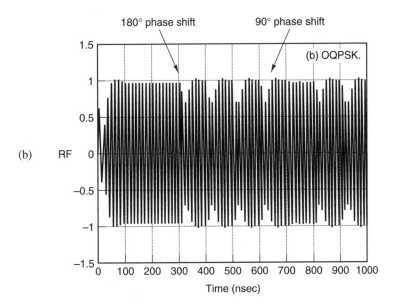

Figure 4–3 Phase transitions at the bit intervals for QPSK(a) and OQPSK(b).

both values, $a_I(t)$ and $a_Q(t)$, will produce a 180° phase shift and a zero envelope at that instant.

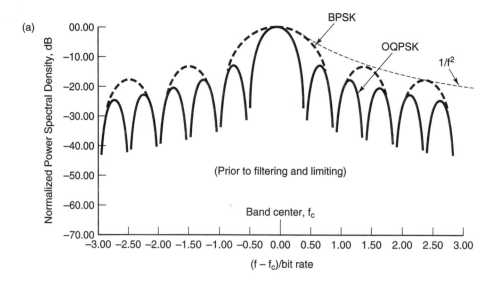

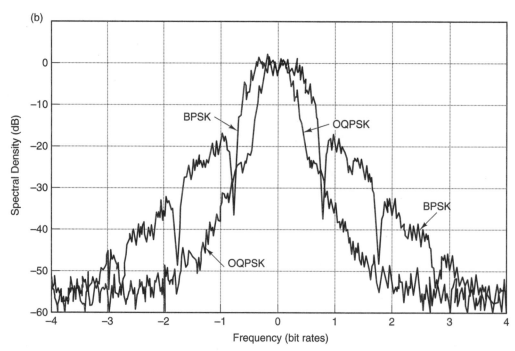

Figure 4–4 Spectral sidelobe regrowth of BPSK and absence of regrowth of OQPSK after hard-limiting amplification.

Linear Modulation

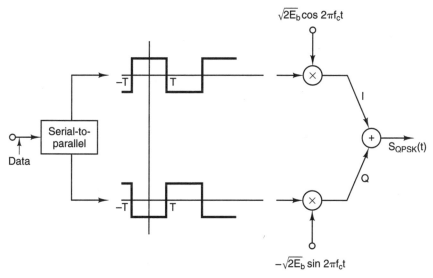

(a) QPSK quadrature modulation generation.

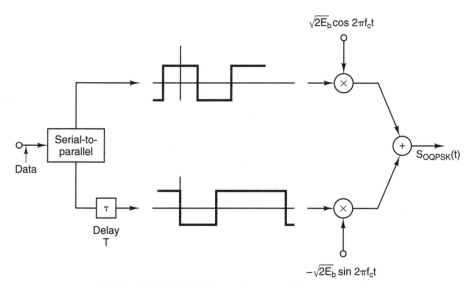

(b) Offset QPSK (OQPSK) quadrature modulation generation.

Figure 4–5 Generation of QPSK and OQPSK modulation waveforms.

In OQPSK generation, the bit display is similar to that of QPSK, but the Q channel sequence is displaced by one bit. This is shown in Figure 4–5(b). This prevents a 180° change since the bits in the two channels cannot change simultaneously. Another observation which can be made about OQPSK is that phase transitions occur twice as often as with conventional QPSK. Even though we have displaced one stream by one bit, the composition of the QPSK and OQPSK pulse trains have not changed; therefore, we would expect the power spectral density of QPSK and OQPSK to be identical.

4.3 Continuous Phase Modulation

OQPSK shows further improvement in removing the deleterious impulsive phase changes at the transitions. If we can remove the phase transitions completely, we can generate a continuous phase signal. A waveform with this property, often referred to as minimum shift keying (MSK), may be generated by quadrature modulation similar to that described above for OQPSK. In lieu of using rectangular pulses as in OQPSK, the two binary channels that modulate the I and Q channels of the carrier are shaped by one-half cycles of a sinusoidal waveform. This scheme is illustrated in Figure 4–6(a). The rectangular binary elements are replaced by sinusoidal pulses of the form

$$f(t) = \cos(2\omega t / 4T) \qquad (4.2)$$

Equation (4.2) above therefore becomes

$$s(t)_{MSK} = a_I \cos(\pi t / 2T) \cos(2\pi f_c t) + a_Q \sin(\pi t / 2T) \sin(2\pi f_c t) \qquad (4.3)$$

This signal actually represents a digital FM signal, which is frequency shift keyed, meaning that either a high tone or a low tone is transmitted during each bit interval. The modulation index is $h = 0.5$ and the carrier is not transmitted.

The terms $(\pi t / 2T)$ are the information phase terms associated with the waveform. Notice that the phase is a linear function of time, and at $t = T$, the phase has undergone a 90° shift during a symbol interval. To be strictly correct, the total phase change is

$$\phi(t)' = 2\pi f_c + (\pi t / 2T) + \phi_o(t) \qquad (4.4)$$

where $2\pi f_c t$: is the increasing carrier phase with time, $\pi t / 2T$ is the information-carrying phase, which has occasionally been referred to as excess phase (over and above that offered by the carrier phase term), and $\phi_o(t)$ is the phase angle of the carrier at the start of the bit interval. This may be equal to 0° or have a finite value.

The phase of the waveform is advanced or retarded precisely 90° with respect to the linearly increasing carrier phase.

Figure 4–6(b) defines the components of the MSK signal defined by Equation (4.3). The combined signals from the I and Q channels are also shown in Figure 4–6(b).

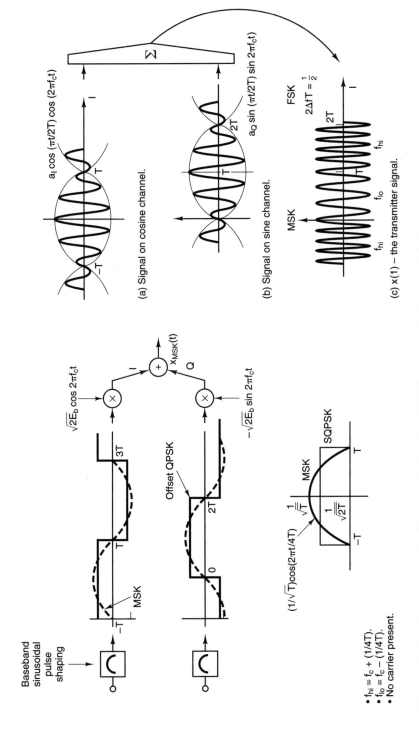

Figure 4–6 Generation of minimum shift keying (MSK) via quadrature modulation.

It is observed that the phase is continuous at the transitions, and the two frequencies are displaced with the high frequency equal to $f_c + (1/4T)$ and low frequency or tone by $f_c - (1/4T)$. Once again, this is for a digital FM signal with a modulation index equal to $h = 0.5$.

MSK is a special type of continuous phase frequency shift keying (CPFSK), often referred to as fast frequency shift keying (FFSK), where the peak frequency deviation, Δf, is equal to one-quarter the bit rate and a modulation index of $h = 0.5$. In digital FM notation:

$$h = 2\Delta f / R_b \tag{4.5}$$

where $2\Delta f$ is the peak to peak RF frequency shift. Therefore, for $h = 0.5$, the spectral tones above and below the carrier are displayed as

$$\Delta f = R_b / 4 = 1 / 4T_b \tag{4.6}$$

We should not, however, infer from this that the spectrum of the transmitted signal consists of impulses of frequencies, $f_c \pm (1/4T)$.

The transmitted tones are either one-quarter data rate $(1/4T)$ above or one-quarter data rate below the carrier. This is the minimum spacing with which signal orthogonality over the symbol interval can be achieved, thus the name MSK (see Section 4.6).

The power spectral density comparison of OQPSK and MSK is shown in Figure 4–7. The sidelobes of MSK fall off at a greater rate than OQPSK, but it has a wider main lobe. The main lobe is confined to $f_c \pm (3R_b/4)$, translated to RF, and 99% of the power is located in the main lobe. For QPSK or OQPSK, the main lobe is bounded by $f_c \pm (R_b/2)$, and contains 92% of the power.

Some of the attributes of MSK modulation are

- The phase across the bit transitions is continuous with no abrupt phase changes. The phase is piecewise-continuous, but the derivative of the phase is still discontinuous.
- Phase shift changes linearly during a pulse interval and has a maximum value of 90°.
- Minimum frequency separation between data bits still maintain signal orthogonality. (See Figure 4–8 for graphical details.)
- A constant envelope is maintained after filtering and Class-C amplification. That is, the AM/AM and AM/PM nonlinearities do not affect the restored uniform envelope (AM/AM helps), or restore the spectral sidelobes.

An alternate method of generating an MSK waveform is via direct frequency modulation. Figure 4–9 shows a simplified block diagram of a direct FM modulator. This is a special case of continuous frequency shift keying (CFSK) [4].

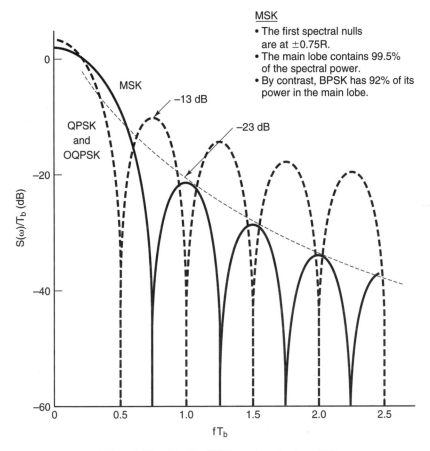

Figure 4-7 Power spectral density of QPSK, OQPSK, and MSK.

The filter, $G(f)$, is the premodulation filter which shapes the input over a finite interval, $0 \leq t \leq LT$, where L is the length of the pulse (per unit T). The shape of $G(f)$ determines the smoothness of the transmitted information-carrying phase. The $G(f)$ output, $g(t)$, multiplied by $2\pi h$ is used to FM-modulate the VCO, thus producing a CPM signal output. The impulse response of $G(f)$ is an NRZ sequence of rectangular pulses.

The output of the FM modulator can be given as

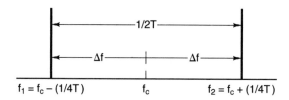

Closest tone spacing in MSK(h) to still maintain signal orthogonality is
$h = \Delta f/R_b = (f_2 - f_1)/R_b = 0.5$

From the figure above, the frequency deviation of the VCO is
$\Delta\omega = 2\pi h g(t)$
but $g(t) = 1/2T$ and $h = 1/2$
$\therefore \Delta\omega = 2\pi(1/2)(1/2T) = \pi/2T$
or $\Delta f = 1/4T = R_b/4$

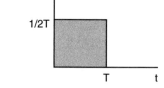

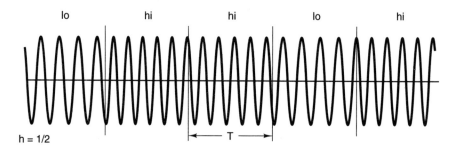

- Note there are no phase jumps at the bit transitions like in linear modulation waveforms.

Figure 4–8 Constant envelope MSK waveform.

$$s(t)_{\text{CPM}} = \sqrt{2E/T} \cos\left[2\pi f_c t + 2\pi h \int_0^t \sum_{n=0}^{\infty} a_n g(\tau - nT) d\tau\right] \quad (4.7)$$

The symbols, a_n, are equally likely and independent where the second term in the square brackets carries the information phase, that is

$$\phi(t) = 2\pi h \int_0^t \sum_{n=0}^{\infty} a_n g(\tau - nT) d\tau \quad (4.8)$$

Continuous Phase Modulation

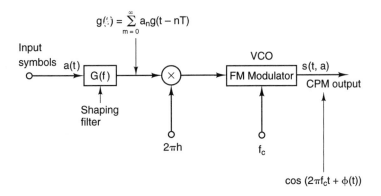

Figure 4-9 Alternate method of generating an MSK waveform.

where E is the energy per symbol, $a_n = \pm 1$, and $g(t)$ is the shaping filter output which determines the smoothness of the transmitted phase information. In this case, $g(t)$ is a digital NRZ data sequence.

The baseband phase pulse is obtained by integrating a frequency pulse, $g(t)$ (note $g(t)$ are rectangular pulses):

$$q(t) \triangleq \int_{-\infty}^{t} g(\tau) d\tau \qquad (4.9)$$

For MSK, a rectangular pulse with amplitude, $1/2T$, and duration, T, $g(t)$, is normalized such that

$$\int g(t) dt = \frac{1}{2} \quad \text{or} \quad q(\infty) = \frac{1}{2} \qquad (4.10)$$

The information-carrying phase may therefore be put in the form

$$\phi(t, \underline{a}) = 2\pi h \sum a_n q(t - nT) \qquad a_n = \pm 1 \qquad (4.11)$$

(a) Full response.

$$g(t) = \begin{cases} g(t) & 0 \leq t \leq T \\ 0 & \text{otherwise} \end{cases}$$

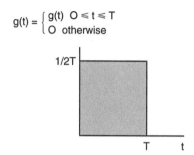

(b) Partial response.

$$g(t) = \begin{cases} g(t) & 0 \leq t \leq LT \\ 0 & \text{otherwise} \end{cases} \quad L > 1$$

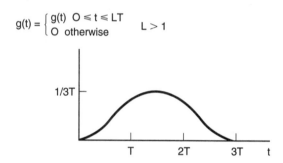

Figure 4–10 Two classes of CPM signals.

As noted previously, if h = 1/2, the total contribution to the phase integral for each pulse, q(t), equals ±p/2. If g(t) is spread over several symbols, g(t) covers several symbols and is referred to as a partial response.

There are two classes of modulating signals. One is the full response, where a T pulse is shaped over one interval. This was previously described (see also Figure 4–10). The other is partial response, where shaping is over several bit intervals. This is also shown in Figure 4–10. In general, if $h = 1/2$, the total contribution to the phase integral, $\phi(t)$, for each pulse, $g(t)$, equals 90°. In the partial response case, this is spread out over several symbols, since $g(t)$ lasts for more than one symbol.

4.4 Phase Trellises in CPM

In previous sections we referred to an information-carrying phase. This phase changes phase during the symbol interval and for CPM signals, the phase is continuous from symbol to symbol with no abrupt changes. Consider, for example, the phase trajectory associated with MSK.

Figure 4–11 shows the rectangular data stream at the output of the filter. The phase trellis for this bitstream is indicated in Figure 4–11(b). In this case, we assume the initial starting phase to be zero degrees. During each bit interval, the phase of the MSK carrier is shifted linearly (±45° slope) with time by ±π/2 radians. The phase of the MSK waveform is advanced or retarded precisely 90° with respect to the carrier depending on the bit value. As shown in the figure, selection of the low tone (0, or first bit) causes a retardation of the phase by π/2 during the bit interval. A high tone (second bit) causes an advance of the phase by 90° (here the phase moves upward at 45°). From our previous analysis, the phase response, $q(t)$, is obtained by integrating a frequency pulse, $g(t)$. The trellis curve is piece-wise linear because the data sequence of pulses, $g(t)$, is rectangular. Also note that we have full response signaling as shown in Figure 4–10.

Figure 4–11(c) shows that for $a_n = +1$, there is an increase in frequency output above the carrier to $f_c + \Delta f = f_c + (1/4T)$, or an advance of phase by π/2. For $a_n = -1$, the frequency becomes $f_c - \Delta f = f_c - (1/4t)$, or a retardation of the signal phase. The unmodulated carrier is not transmitted.

We have observed in the phase trellis for MSK that the curve is piece-wise continuous in time without discrete phase changes. Gradual phase trajectories may be obtained by using shaping pulses which are smoother. More will be explained on this subject in a later section, when we discuss Gaussian MSK (GMSK).

Another interesting interpretation of the phase trajectory of MSK in signal space (I-Q diagram) is represented in Figure 4–11(d) [18]. If we start at the epoch point on the circle at the right, the zig-zag traces reflect the phase changes during each bit interval. For example, for a phase change of the first bit, the phase changes by 90° and the trace moves clockwise. The next bit provides a plus phase of 90°. Here there is a reversal of the trace and it moves counter-clockwise back to the I axis. The phase trace moves back and forth until all 16 bits have been accounted for.

It is interesting to make a few comments on the signal space diagram for a CPM signal, which is depicted in Figure 4–12. For a CPFSK signal, we can represent a CPM signal in signal space by a circle, where the points on the circle correspond to the combined amplitude and phase of the carrier. Constellations for BPSK, QPSK, OQPSK, and π/4-QPSK occupy particular points on the plane. This is not the case for CPM signals. There are no set positions where the symbols reside because the phase of the carrier is time-variant. The migration of the symbols may move in either direction, and the symbols are never at rest.

On the circle, the differential phase between bit intervals depends on the modulation index. For increasing values of h, the phase variation between bit intervals becomes increasingly larger, in addition to increasing the bandwidth. For example, for $h = 1$, the frequencies are farther apart than for $h = 1/2$.

Virtual locations (no terminal phase states) on the signal state space diagram are represented in Figure 4–13 (locations are arbitrary, but uniform). The deployment reveals the linear phase shift (±) during the transition interval. Notice that for $h = 1$, the phase variations are 180°, as compared to, say, $h = 1/2$, where variations are 90°.

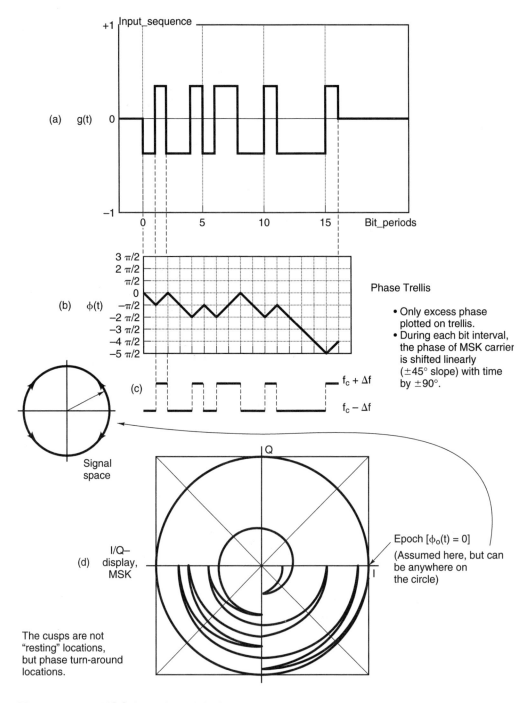

Figure 4–11 VCO-based modulation.

GMSK Modulation

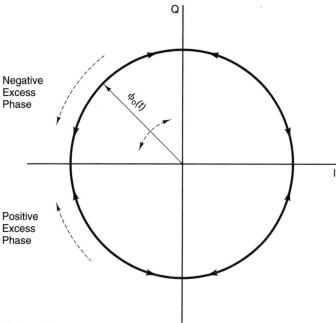

- No discrete symbol positions are manifested.
- No resting locations.

After turning through an angle, when a symbol is transmitted, it does not remain at rest, but continues in the same direction if the symbol is repeated, or reverses direction if the opposite signal is transmitted.

Clearly, this is unlike BPSK, QPSK, and $\pi/4$QPSK, where discrete points are located on the plane.

Figure 4–12 Signal state space diagram for a CPM signal.

4.5 GMSK Modulation

GMSK is gaining in popularity as a modulation scheme for several wireless communications applications. In mobiles, power is at a premium and the use of CPM-type signals permits the use of efficient Class-C amplifiers. GMSK is used in the European GMS cellular standard and also in the American PCS-1900 standard.

In previous sections, we have indeed shown that MSK modulation produces a continuous phase across bit transitions with no abrupt change in phase. Therefore, limiting band to satisfy out-of-band emission requirements will not lead to envelope fluctuations. However, there are piece-wise discontinuities which have a tendency to produce sidelobes and a relatively broad main lobe. Both of these areas can be improved by

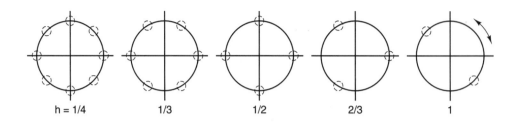

h	Δφ	BW
1/4	45°	
1/3	60	
1/2	90	Increasing
2/3	120	
1	180	

Figure 4–13 Phase variations for several values of h.

using premodulation shaping filters, which smooth out phase excursions, reduce additional sidelobes, and narrow main lobes. One approach in this direction is to use a Gaussian shaping filter.

Gaussian Minimum Shift Keying (GMSK) is a continuous phase modulation scheme generated by filtering NRZ data with a Gaussian shaping filter, as shown in Figure 4–14. An NRZ sequence drives a Gaussian LPF with the Gaussian shaped output used to FM modulate a VCO. A GMSK-modulated waveform is characterized by the BT product, where B is the bandwidth of the filter and T is the duration of the data bit. Unlike MSK, which uses full-response signaling, GMSK uses partial-response (see Figure 4–10), where the signal is spread over several T periods. For example, for $BT = 0.3$, the Gaussian shaping filter output, $g(t)$, extends over three bit intervals.

The impulse response of the Gaussian filter for several values of BT is displayed in Figure 4–15(a). Additional information on the Gaussian LPF can be found in Appendix A. Note that as BT decreases, $g(t)$ spreads out (also creating ISI). However, this is tempered by the fact that a small BT generates a narrowband signal. It is recalled from signal theory that the system functions, $H(w)$ and $h(t)$, are Fourier Transform pairs.

GMSK Modulation

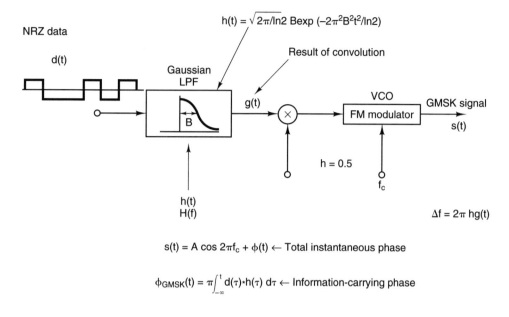

Figure 4–14 Generation of Gaussian MSK.

The integral equation in Figure 4–14 gives the phase trajectory during the symbol interval. It is recalled from the MSK equation cited previously, that the integral of a similar equation gave a ramping phase over the bit intervals. Here, the phase is shaped like an ogive curve since we are integrating a $g(t)$ with a Gaussian shape. See, for example, Figure 4–15(b) for $BT = 0.3$. The observant reader may notice that the full response of MSK is converted to a partial-response scheme after Gaussian filtering.

For GMSK modulation, both the spectral sidelobes' power and the width of the main lobe can be reduced, depending on the value of BT. The power spectral density plotted as a function of the normalized frequency difference from the carrier center frequency is plotted in Figure 4–16 [17]. Here, B_b is the bandwidth of the Gaussian low-pass filter and T is the bit period. The spectrum main lobe becomes increasingly narrower and the sidelobes are reduced as BT decreases. This further reduces the discontinuous phasing manifested by MSK modulation. The spectral sidelobes practically disappear for small values of B_bT. However, it should be realized that reducing B_bT results in a more compact power density spectrum bit.

Ch. 4 • Dynamics of Linear and Continuous Phase Modulation Methods in Digital Communications

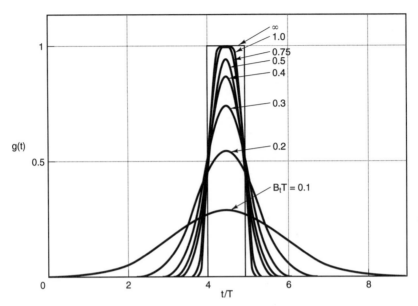

(a) Pulse response of a Gaussian filter.

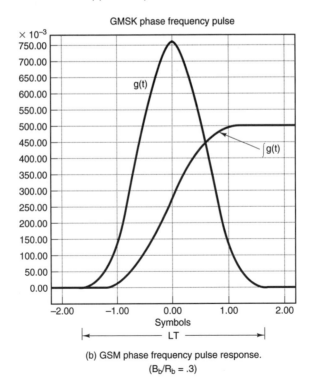

(b) GSM phase frequency pulse response.
$(B_b/R_b = .3)$

Figure 4–15 Impulse response of Gaussian filters for parametric values of *BT*.

GMSK Modulation

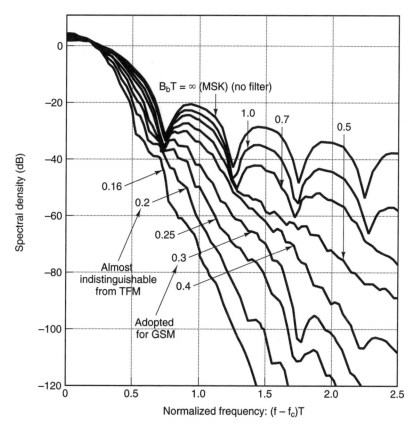

Figure 4–16 Power spectra of GMSK. Copyright © 1981 by the Institute of Electrical and Electronics Engineers, Inc. Reprinted with permission.

The time domain introduces ISI, which can be demonstrated in eye diagrams by the eye closing up. Therefore, in practice, there is a trade-off between B_bT and the amount of tolerable interference. Investigators have found that the degradation is not severe if the 3 dB bandwidth bit duration product (B_bT) of the filter is greater than 0.5 [21]. It is also noticed that as the spectrum becomes narrower, there is less energy in the signal and thus E_b/N_o falls off, increasing the BER.

Some practical test spectra of GMSK for various values of BT are shown in Figure 4–17. Figure 4–17(a) shows the power spectrum of three contiguous channels using MSK modulation. As can be observed, the interference from an adjacent channel at the center of the central channel is down about 37 dB. The vertical grid is 10 dB/division. In Figure 4–17(b), a Gaussian shaping filter is used with a $BT = 0.5$. The center channel shows a flanking signal suppression of about 60 dB. In Figure 4–17(c), the Gaussian filter is narrowed further with $BT = 0.25$. Here, interference at the central channel is unmeasurable, and the suppression at the extremes of the channel skirt is about 55 dB.

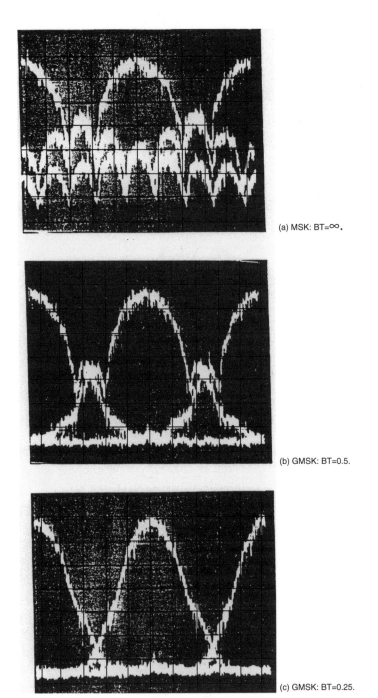

Figure 4–17 Practical power spectra for GMSK for particular values of BT. From *Microwaves & RF* (August 1984). Reprinted with the permission of the publisher.

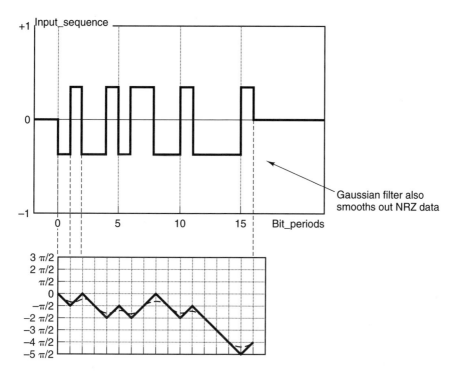

- Solid trajectory on trellis: MSK modulation.
- Dotted line on trellis: GMSK modulation depicting smoothing action of Gaussian filter.

Note: The phase trajectory of GMSK is a "string" of contiguous, confluent, ogive-shaped curves (as shown in Figure 4.15), resulting from integrating the g(t) function.

Figure 4–18 Smoothing out of the phase trajectory by pre-modulation Gaussian filtering.

Figure 4–11 showed the phase history for a data stream generating MSK. Each bit change resulted in a phase change of 90° and was piecewise-continuous at the bit transitions. Figure 4–18 shows the same bitstream, as well as the MSK phase trellis. The phase changes for GMSK are superimposed. Here, however, the sharp corners on the phase curve have been smoothed out by Gaussian filtering. The shape of $g(t)$ determines the smoothness of the transmitted information-carrying phase, and the phase response, $q(t)$, is obtained by integrating a frequency pulse, $g(t)$. The phases at the transitions are less abrupt, which results in reduction of the spectral sidelobes. We further notice that the phase excursions during a bit interval fall short of ±90°. In addition to the phase tra-

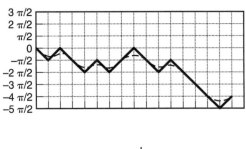

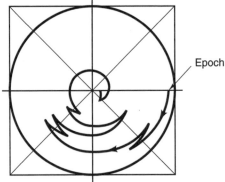

Note: The first cusp corresponds to the first phase turnover of the first bit.

Figure 4–19 Smoothing of the phase trajectory by the action of the Gaussian low-pass filter.

jectory on the phase trellis, the movement may also be displayed on a signal state space diagram, or I-Q plane. This is shown in Figure 4–19. The Gaussian phase curve and I-Q plane display are correlative. We leave this as an exercise for the reader.

4.6 Tamed Frequency Modulation (TFM)

An additional improvement in the phase variations can be achieved by using tamed FM (TFM). In this scheme, data is coded correlatively so that fewer and smaller phase variations are realized [8]. The authors in the cited references propose that a change in phase by 90° occurs if the three succeeding bits have the same polarity. If no phase change takes place, the bits have zig-zag polarity. This is demonstrated in Figure 4–20, which was taken from the reference indicated.

On the signal state space diagram, the phase variations are reduced and the phase trajectory is smoother. Also, the phase difference between bits corresponds to $h = 1/2$, or a phase variation of 45°. The power spectrum of TFM is roughly that corresponding

Signal Orthogonality

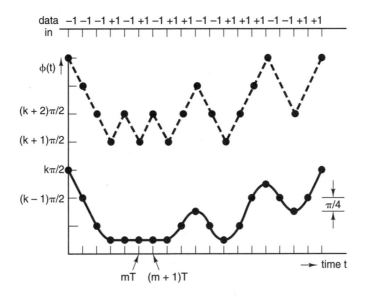

Figure 4-20 Phase behavior of MSK (- - -) and TFM (——). Copyright © 1978 by the Institute of Electrical and Electronics Engineers, Inc. Reprinted with permission.

to $B_bT = 0.2$, as shown in Figure 4-16. Clearly, the spectral sidelobes are practically nonexistent.

4.7 Signal Orthogonality

Signal orthogonality is achieved if the correlation coefficient between two signals equals zero.

$$\rho = \int_0^T s_1(t) \cdot s_0(t) dt = 0 \quad (4.12)$$

If we assume the signalling states are normalized to unity energy, they can be represented by

$$s_1(t) = \sqrt{2/T} \cos(2\pi f_1 t + \phi_1)^{[1]} \qquad 0 \le t \le T \quad (4.13)$$

$$s_2(t) = \sqrt{2/T} \cos(2\pi f_0 t + \phi_2) \qquad 0 \le t \le T \quad (4.14)$$

---

1. ϕ_1 and ϕ_2 are arbitrary phase angles that may or may not be equal to zero.

Equation (4.12) then becomes

$$\rho = \int_0^T (\sqrt{2/T}\cos 2\pi f_1 t \cdot \sqrt{2/T}\cos 2\pi f_0 t)\, dt \tag{4.15}$$

Using the trigonometric identity [23]

$$\cos x \cos y = (1/2)\cos(x-y) + (1/2)\cos(x+y) \tag{4.16}$$

and plugging in Equation (4.15), we obtain

$$\rho = \int_0^T [(1/T)\cos(2\pi f_1 - 2\pi f_0)t + \cos(2\pi f_1 + 2\pi f_0)t]\, dt \tag{4.17}$$

Integrating and plugging in the limits, we obtain

$$\begin{aligned}\rho &= (1/T)(1/2\pi(f_1-f_0))\sin(2\pi(f_1-f_0)t)\big|_0^T \\ &\quad + (1/T)(1/2\pi(f_1+f_0))\sin(2\pi(f_1+f_0)t)\big|_0^T \\ &= \sin[2\pi(f_1-f_0)T]/2\pi(f_1-f_0)T + \sin[2\pi(f_1+f_0)T]/2\pi(f_1+f_0)T\end{aligned} \tag{4.18}$$

If $f_1 + f_0 \gg f_1 - f$, Equation (4.18) reduces to

$$= \sin[2\pi(f_1-f_0)T]/2\pi(f_1-f_0)T = \sin 2\pi\Delta f T / 2\pi\Delta f T \tag{4.19}$$

where $\Delta f = f_1 - f_0$ is the separation between high and low frequencies.

Equation (4.19) is shown plotted in Figure 4–21, which is recognized to be of the form $\sin x/x$. We notice that when $f_1 = f_2$, $\rho = 1$. In addition, the second term in Equation (4.18) has a null value since the frequency deviation is much larger than the data rate, $\Delta f \gg 1/T$.

We notice that when $\Delta f = 1/2T, 1/T, 3/2T, ..., n/2T$ the function goes to zero and the correlation equals zero. The case where $\Delta fT = \Delta f / R_b = 1/2$ is for MSK, which is the closest frequency separation for orthogonality.

It can be shown that the optimum probability of error of FSK occurs when the correlation function is negative. In this display, this is when $\rho = -0.212$. This results from the probability of error relationship for equal energy signals.

$$\begin{aligned}P(e) &= Q(\sqrt{(E/N_o)(1-\rho)}) \\ &= Q(\sqrt{1.22(E/N_o)})\end{aligned} \tag{4.20}$$

Therefore, the optimum FSK is +0.86 dB better than orthogonal FSK.

Signal Orthogonality

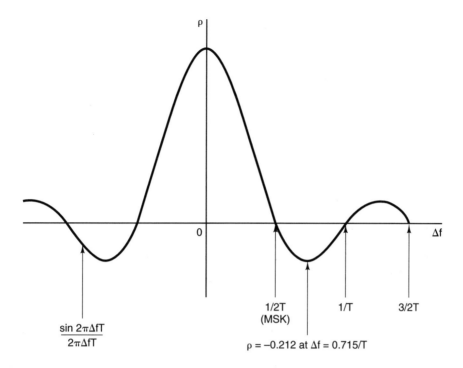

Figure 4–21 Correlation showing regions of signal orthagonality.

Mathematically, the degree of difference between two signals in signal space can be expressed in terms of their Euclidean distance (or via the correlation factor indicated above), which is given by the square of the distance between the two signals.

$$d^2 = \int_0^T [x(t) - y(t)]^2 \, dt \qquad (4.21)$$

where

$$x(t) = \sqrt{E_1} s_1(t), \; y(t) = \sqrt{E_0} s_0(t) \qquad (4.22)$$

Note that

$$\int_0^T s_1^2(t) \, dt = \int_0^T s_0^2(t) = 1 \qquad (4.23)$$

$$\therefore d^2 = \int [\sqrt{E_1}s_1^2(t) + \sqrt{E_0}s_0^2(t) - 2\sqrt{E_1E_0}s_1(t)s_0(t)]dt \qquad (4.24)$$

$$= E_1 + E_0 - 2\sqrt{E_1E_0}\int_0^T s_1(t)s_0(t)dt$$

$$= E_1 + E_0 - 2\rho\sqrt{E_1E_0}$$

where the integral term equals the correlation coefficient.

For equal energy signals, $E_1 = E_0$,

$$d^2 = 2E(1-\rho) \qquad (4.25)$$

To achieve the maximum value of d, or to realize best performance in an AWGN environment, ρ should be equal to -1. These signals are defined as antipodal signals. The best possible separation is therefore

$$d = 2\sqrt{E} \qquad (4.26)$$

BPSK modulations are antipodal signals and in the presence of AWGN provide the best performance. They are therefore used as a benchmark for other modulation types.

It can be shown that for any pair of *binary signals* (equal or unequal), $P(e)$ only depends on the distance, d, between them and not their locations.

$$P(e) = Q(d/\sqrt{2N_0}) \qquad (4.27)$$

where Q is

$$Q(x) = 1/\sqrt{2\pi} = \int_x^\infty \exp(-y^2/2)dy \qquad (4.28)$$

(not a closed form expression)

If we plug d from Equation (4.24) into Equation (4.27), we obtain

$$P(e) = Q[\sqrt{E_1 + E_0 - 2\rho\sqrt{E_1E_0}/2N_0} \qquad (4.29)$$

Signal Orthogonality

if $E_1 = E_0$

$$P(e) = Q(\sqrt{(E/N_o)(1-\rho)}) \qquad -1 \leq \rho \leq 1 \qquad (4.30)$$

if $\rho = -1$

$$P(e) = Q\sqrt{2E/N_o} \qquad (4.31)$$

This is the $P(e)$ for antipodal signals of which BPSK ia a form.

For a four-point QPSK constellation, there are two Euclidean distances, that is,

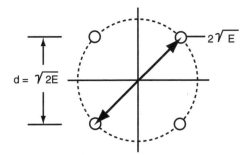

Here, smaller Euclidean distances are manifested, thus making this modulation (for example) more susceptible to noise. However, the bandwidth efficiency is better (i.e., 2 b/s-Hz vs. 1 b/s-Hz for BPSK). Similarly, for M-ary PSK with M greater than 2, the Euclidean distance becomes progressively smaller. For MPSK, the minimum distance is given as

$$d_{\text{mim}} = 2\sqrt{E_s}\sin(\pi/M) \qquad (4.32)$$

where $E_s = E_b \ln M$ and $[\log_2 M = \log_{10} M / \log_2 2]$ = Bits/Symbols

The probability of symbol error for M-ary signals is given by

$$P(e) \leq (M-1)Q(d_{\min}/\sqrt{2N_o}) \qquad (4.33)$$

For *orthogonal* signals, we have a different situation concerning the Euclidean distance. We can easily project this to be a CPM signal for $h = 0.5$ (orthogonal condition). For the *virtual* symbol display in Figure 4–13, for $h = 0.5$, we note the Euclidean distance, realizing that at $\rho = 0$ we have

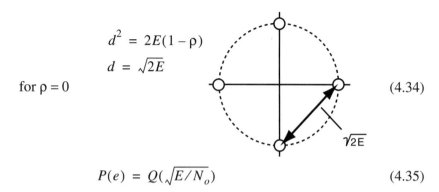

$$d^2 = 2E(1-\rho)$$
$$d = \sqrt{2E}$$
for $\rho = 0$ \hfill (4.34)

$$P(e) = Q(\sqrt{E/N_o}) \tag{4.35}$$

The probability of error for binary orthogonal signals is 3 dB worse than for the antipodal case. Therefore, we must increase energy by 3 dB to realize the same BER performance as with BPSK.

4.8 Conclusions

Linear modulations are popular waveforms that are used extensively in both terrestrial wireless and satellite systems. The advantage is that they have a relatively higher spectral efficiency as compared with the more recent CPM types. On the down side, linear modulation types perform best in linear channels where band-pass filtering is used (to remove out-of-band emissions). The requirement to use Class-A amplifiers prevents the onset of spectrum regrowth. CPM signals, on the other hand, are constant envelope (after filtering) and can be amplified by more efficient Class-C amplifiers. This technology is particularly applicable in cellular systems, where the hand transceivers are limited in power and battery weight. However, where spectral efficiency is of paramount importance, it may be necessary to go with linear modulation (e.g., π/4-QPSK), which has a theoretical efficiency of 2 b/s-Hz (practically, about 1.6 b/s-Hz). Higher order linear modulation types have greater efficiency (M-ary PSK, M-ary QAM), but at the expense of greater E_b/N_o to achieve the desired BER.

Spectral efficiencies of constant envelope modulation types are not as high and generally less than those found in linear modulation. However, to say that linear modulation systems offer higher spectral efficiencies should be qualified. A proviso is that linear systems use linear amplification, after filtering, to prevent deleterious spectrum regeneration. Figure 4–22 demonstrates the filtered spectrum of QPSK in comparison to CPM modulations. Efficiency in wireless CPM systems runs between 1 and 1.6 b/s-Hz. Therefore, where spectral efficiency is important, it may be necessary to foresake CPM and use linear modulation schemes, even though the system will take a "hit" on the additional power and battery weight penalties. For example, π/4-QPSK has been used in several cellular system handsets. One attempt to improve its performance is the use of negative feedback control to suppress the sidelobes [22].

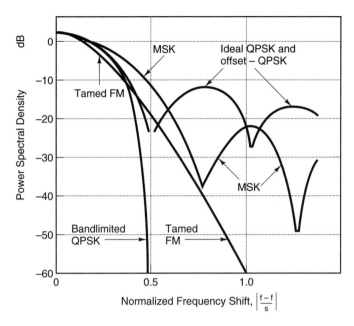

Figure 4–22 Spectra of various modulations.

In Chapter 5 we will deal with channel coding. That is, introducing redundancy to the data bits in order to provide a means of detecting and/or correcting transmission errors. We will discuss forward error correction (FEC) in which re-transmission is not necessary in order to ascertain the error(s) in transmission. This differs from automatic repeat request (ARQ) where re-transmission is required because of an error in transmission. There are basically two main types of FEC; that is, Block Coding and Convolutional Coding. We will elucidate on these in Chapter 5.

4.9 References

[1] S. Pasuppathy, "Minimum Shift Keying: Spectrally Efficient Modulation," *IEEE Communications Magazine*, July 1979.

[2] I. Kalet, "Modern Digital Communications Techniques," Course given at George Washington University, August 1994.

[3] S. A. Gronemeyer and A. L. Mc Bride, "MSK and Offset QPSK Modulation," *IEEE Trans. Communications*, August 1976.

[4] C. W. Sundberg, "Continuous Phase Modulation," *IEEE Communications Magazine*, April 1986.

[5] J. B. Anderson and C. W. Sundberg, "Advances in Constant Envelope Coded Modulation," *IEEE Communications Magazine*, December 1991.

[6] L. E. Larson, *RF and Microwave Circuit Design for Wireless Communications*, Artech House Pub., Boston, MA, 1996.

[7] C. R. Ryan, et al., "760 Mbit Serial MSK Microwave Modem," *IEEE Trans. Communications*, 1980.

[8] F. de Jager and C. B. Dekker, "Tamed Frequency Modulation – A Novel Method to Achieve Spectrum Economy in Digital Transmission," *IEEE Trans. Communications*, May 1978.

[9] R. de Buda, "Coherent Demodulation of Frequency Shift Keying With Low Deviation Ratio," *IEEE Trans. Communications*, June 1972.

[10] N. W. R. Bennett and S. O. Rice, "Spectral Density and Autocorrelation Functions Associated With Binary Frequency Shift Keying," *Bell Sys. Tech. Journal*, Vol. XLII, September 1963.

[11] N. Rydbeck and C. E. Sundberg, "Recent Results on Spectrally Efficient Constant Envelope Digital Modulation Methods," *Proc. ICC*, Boston, MA, June 1979.

[12] W. J. Weber, P. H. Stanton, J. T. Sumida, "A Bandwidth Compressive Modulation System Using Multiple Amplitude Minimum Shift Keying (MAMSK)," *IEEE Trans. Communications*, May 1978.

[13] F. Amoroso, "Pulse & Spectrum Manipulation in the Minimum (Frequency) Shift Keying (MSK) Format," *IEEE Trans. Communications*, May 1976.

[14] M. K. Simon, "A Generalization of Minimum Shift Keying (MSK)—Type Signaling Based Upon Input Data Symbol Pulse Shaping," *IEEE Trans. Communications*, Aug. 1976.

[15] M. Rabzal and S. Pasupathy, "Spectral Shaping in Minimum Shift Keying (MSK) Type Signals," *IEEE Trans. Communications*, January 1978.

[16] J. B. Anderson and D.P. Taylor, "A Bandwidth Efficient Class of Signal Space Codes," *IEEE Trans. Information Theory*, November 1978.

[17] B. Pattan, *Satellite-Based Global Cellular Communications*, McGraw-Hill Co., New York, 1998.

[18] British Electrical Engineering Magazine, date unknown.

[19] M. Ishizuka and K. Hirade, "Optimum Gaussian Filter and Deviated-Frequency Locking Scheme for Coherent Detection of MSK," *IEEE Trans. Communications, June 1980.*

References

[20] T. Aulin, et al., "Continuous Phase Modulation-Part II: Partial Response Signaling," *IEEE Trans. Communications*, March 1981.

[21] K. Murota and K. Hirade, "GMSK Modulation for Digital Mobile Radio Telephony," *IEEE Trans. Communications*, July 1981.

[22] Y. Akaiwa and Y. Nagata, "Highly Efficient Digital Mobile Communications With a Linear Modulation Method," *IEEE J. Sel. Areas in Communications*, June 1987.

[23] R. S. Burington, *Handbook of Mathematical Tables and Formulas*, Handbook Pub., Sandusy, OH, 1958.

CHAPTER 5

Error Control Coding

5.1 Introduction

There is a growing need for reliable transmission of high-quality voice and digital data over the terrestrial and satellite-based wireless systems. These systems must be limited in power and spectrum requirements, with the latter prevailing.

To lessen the spectrum requirements, spectrum- and bandwidth-efficient modulation schemes are evolving, including CPM, MPSK, and QAM. In applications where power is at a premium, coding schemes are used. In fact, most terrestrial and satellite systems use some form of coding to conserve power. However, on the downside, coding usually implies that additional bandwidth is required. Fortunately, this situation has been somewhat ameliorated by new concepts such as trellis coded modulation (TCM), where bandwidth expansion is avoided, but coding gain is achieved. We will not discuss TCM in this chapter, but will confine our attention to traditional coding, where modulation and coding are treated as separate operations in overall system design, that is, modulation with maximally separated signals and coding (block or convolutional) with maximized minimum Hamming distance. TCM will be described in detail in Chapter 6.

The efficiency of a communication system is usually measured by the received energy per bit to noise power density ratio (E_b / N_o) required to achieve a particular BER. This can be given by the relationship.[1]

---

1. Note that this is the fundamental relationship in the absence of coding. If coding is present, the term R is the symbol rate, where $R_s = (R_{biofo}/mr)$, $m = \log_2 M$-ary symbols, and r = code rate. For example, for BPSK, $m = 1$ and no coding, $(r = 1)$, Equation (4.1) applies. If $r \neq 1$ and $m > 1$, we have $E_s/N_o = (P_r/N_o)/R_s$ and $(E_b/N_o) = (P_r/N_o)/Rb$. Since $R_s = R_b/mr$, we obtain $E_b/N_o = (E_s/N_o)(1/mr)$.

$$P_r/N_o = (E_b/N_o)RM \tag{5.1}$$

where P_r/N_o is the received power to noise density ratio given by the familiar link equation

$$P_r/N_o = P_t/G_t(\lambda/4\pi r)^2 L_{\text{misc}}(G_r/kT_s)$$

$$(P_r/N_o)_{\text{dB}} = \text{EIRP} + G/T_s - 20\log(4\pi/\lambda) - k \tag{5.2}$$

where $N_o = kT_s$, the one-sided power spectral noise density, E_b/N_o is the energy per bit/N_o required to achieve a specified system BER, R is the information rate in bits per second, P_tG_t is the transmitted power × transmit antenna gain, G_r is the receive antenna gain, λ is the operating wavelength, and r is the effective range from transmitter to receiver. For a satellite network, r is the slant; for a terrestrial system, the exponent on r will differ from 2 because of the frequent non-line-of-sight propagation.

It is therefore suggested from Equation (5.1) that efficiency results from using a modulation technique which requires the minimum E_b/N_o to realize the desired BER and/or to use coding techniques which reduce the E_b/N_o required for a given BER.

It is of interest to note Shannon's bound in relationship to this discussion, which indicates that the channel capacity for a Gaussian channel with infinite bandwidth and average power, P, is

$$C_\infty = (P/N_o)\ln 2 \text{ bps} \tag{5.3}$$

From Equation (5.1), if $R = C_\infty$, we obtain from (5.1) and (5.2)

$$E_b/N_o = \ln 2 = -1.6 \text{ dB} \tag{5.4}$$

5.2 Code Families

There are two basic categories of error control: 1) Request for Repeat (ARQ), and 2) Forward Error Correction (FEC). For ARQ, redundancy is added prior to transmission. Redundancy at the receiver is used to detect errors, but not to correct them. If errors are detected, a request is made to repeat the message. This therefore suggests that a return to sender path is required to request retransmission for ARQ.

FEC is both an error detection and error correction scheme. Redundant bits are added to the information bits at the encoder. Here, redundancy is used at the receiver decoder to correct errors caused by transmission channel impairments. No return path is necessary as in ARQ. FEC codes are basically of two kinds: 1) block codes, and 2) convolutional codes. Block codes work with symbols (which contain data plus parity bits), and convolutional codes work with individual bits in the information bit-parity-bit sequence.

In block coding, the n-bit output of the encoder depends only on the corresponding k-bit information sequence and is independent of the other k-bit sequences. This system is often termed "without memory" or "memoryless". The encoder memorizes just one data word. The decoder memorizes just one code word. In convolutional encoders, on the other hand, each encoder output sequence depends not only on the current output message, but also on a number of past messages. This system is often termed "with memory".

The most commonly used block codes are as follows:

- Golay.
- Bose, Chaudhuri, Hocquenghem (BCH).
- Hamming.
- Reed-Solomon (R-S).

The encoder for a block code divides the information sequence into information blocks of k-bits. The encoder transforms each k-bit information sequence independently into an n code word. The block code therefore consists of a set of code words of length n referred to as (n, k) block code. The ratio, $r = k / n$, is referred to as the code rate.

In a convolutional encoder, the data stream is connected to an m-stage shift register, followed by n modulo-2 adders which are connected to some registers and then to a commutator which scans the output of the modulo-2 adders to produce the coded word. The data words are intermingled with the parity check bits. Constraint length, K, and rate, r, are normally associated with convolutional encoders. K is normally dictated by the number of registers and a larger K gives greater coding gains, but an increase in decoder complexity.

Decoding is more complicated than encoding. The rate indicates the number of redundant bits added to the code word. For example, $r = 1/2$ indicates that one redundant bit is added for one data bit. In addition, the reciprocal of r gives the bandwidth expansions. That is, 2/1 means a factor of two expansion. Both block codes and convolutional codes require bandwidth expansion.

Both block codes and convolutional codes are capable of correcting random errors. Some block codes are also capable of correcting bursty errors where there is a string of contiguous errors. Convolutional codes, with their attendant decoders, do not have bursty error correction capability.

Several major decoding algorithms are used to decode block and convolutional codes:

Block Codes	Convolutional Codes
Massey	Viterbi
Berlekamp-Massey	Sequential (Fano algorithm)
Euclid	Threshold

We will not discuss most of these schemes, in part because of their complexity. They are all useful in particular situations, offering varying degrees of bit error correction and coding gain. Generally, the more powerful the code, the more complex its decoding and the greater the time delay. The attributes of some of theses codes are indicated in Table 5–1.

Table 5–1 Attributes of Coding.

- Pros:
    - Provides coding gain.
    - Improves reliability of communication channel.
    - Reduces cost of hardware design.
- Cons:
    - Requires bandwidth expansion (tempered by TCM).
    - Increases processing delay through the channel.
    - Increases system complexity (tempered by VLSI)

Rate 1/2	Coding Gain	Error Correction	Complexity
BCH	Fair	Random Errors	Moderate
R-S	Good	Random/Bursty	Moderate
Conv.	Fair/Good	Random	High*

* Function of constraint length.

Communication efficiency is normally measured by referring to "waterfall" curves. These are curves depicting the amount of energy per bit over the noise density ratio required to achieve the desired BER. A typical curve for PSK modulation is shown in Figure 5–1. The Figure shows the PSK trajectory for no coding. The potential improvement in coding gain which can be achieved is to the left of the curve, with the lower bound being at the Shannon limit. Coding gain is the E_b/N_o required before coding is subtracted from the E_b/N_o required after coding for the same BER. For example, for PSK with no coding and with a BER = 10^{-5}, the E_b/N_o required is 9.6 dB. The potential improvement which can be achieved is about 11.2 dB. Coding has moved PSK (for example) closer and closer to the Shannon limit with additional improvements resulting from the advent of higher order modulations. Actually, coding and modulation are in conflict as far as bandwidth efficiency is concerned.

FEC is used to improve digital communication efficiency over a noisy channel. This is especially useful where transmitter power is limited and FEC permits achieve-

Code Families

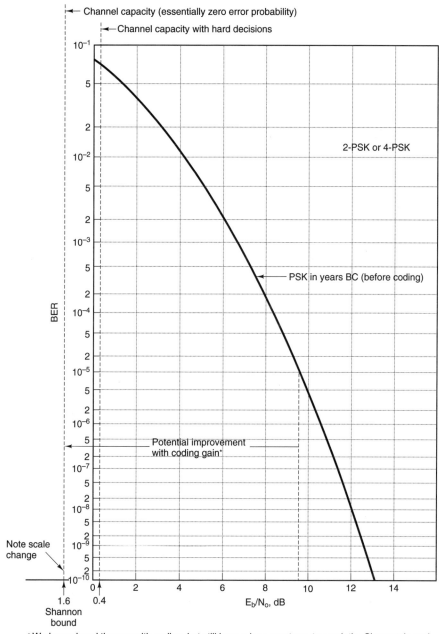

Figure 5–1 Performance curve and potential improvement from coding.

*We have closed the gap with coding, but still have a long way to go to reach the Shannon bound. Migration to the left may be taken as reduced power P, increase in data rate R, or $P/N = (E_b/N_o)R$.

ment of low bit rates at reduced power from an uncoded operation. On the debit side, depending on the code rate, additional bandwidth is required. For a rate 1/2 encoder (a parameter used in coding theory), the system must be able to transmit symbols at twice the data rate, or have a two-fold increase in bandwidth from the uncoded case. The coding rate determines the amount of redundancy added to the data.

FEC therefore reduces the power required to achieve the desired BER, or if you prefer, reduces the BER for the same E_b/N_o. However, there is an increase in bandwidth if the same modulation is used as the uncoded case, or a decrease in throughput if the same bandwidth is used as the uncoded case. Coding gain can be achieved by lowering the code rate (e.g., $r = 1/2 - r = 1/3$), or by increasing code complexity by increasing the constraint length, K. Much higher coding gains can be realized by manipulating the constraint length, K, as opposed to increasing r.

As indicated previously, FEC is concerned with two general classes of codes, namely, block codes and convolutional codes. A third class is occasionally referred to as concatenated codes. These codes are actually the cascading of several codes (usually block-convolutional) to form a more powerful code. The combination is frequently referred to as inner and outer codes. A typical block diagram for concatenated coding is depicted in Figure 5–2. The role of the interleaving block will be explained in a later section.

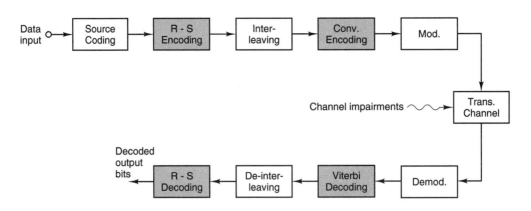

• The codes shown are representative.

Figure 5–2 Signal channel depicting code concatenation with interleaving.

In the block code class, a sequence of source bits is segmented into blocks of k bits and translated into n code bits, with $n - k$ and $p = (n - k)$ redundant bits. The block code is said to have a code rate of $r = k/n$.

The convolutional coder inserts redundant bits without segmenting the source stream into blocks. Instead, it processes the data stream either bit-by-bit or in small groups of bits at one time.

Several advantages accrue from the use of FEC in terrestrial and/or satellite systems. For example, using a convolutional encoder with constraint length $K = 7$, rate $r = 1/2$, with BER = 10^{-5}, and using BPSK or QPSK modulation, the coding gain is

$$E_b / N_o \text{ dB/no coding} - E_b / N_o \text{ dB/with coding} = 5 \text{ dB} \qquad (5.5)$$

This 5-dB improvement would therefore allow

- An equal reduction in transmitter power of 5 dB. In satellite applications, a potential reduction in transponder weight.
- A factor of 3 reduction in antenna area.
- A 5 dB increase in receiver noise temperature.
- A factor of 3 increase in data rate, R, from the relationship

$$(P_{\text{rec}} / N_o)_{\text{rec}} = (E_b / N_o)_{\text{req}} R \qquad (5.6)$$

where $(E_b / N_o)_{\text{req}}$ is the required ratio to achieve BER with coding and P_{rec} / N_o is the ratio required to realize $(E_b / N_o)_{\text{req}}$ *before* coding. P_{rec} / N_o is obtained from the link equation.

Note that for any convolutional code rate, R, the data rate = $R \times$ coded symbol rate, and the following applies:

$$E_s / N_o = Rx(E_b / N_o) \qquad (5.7)$$

where E_b / N_o is the energy per *information bit*-to-noise density ratio, E_s / N_o is the energy per binary *code symbol*-to-noise density ratio, and $P / N_o = E_b / N_o \times$ data rate = $E_s / N_o \times$ coded symbol rate.

5.3 Code Performance

Both the constraint length, K, and code rate, r, have an impact on the performance achieved. The greater the constraint value (with greater increase in the decoder design), the higher the coding gain achieved. A decrease in code rate will also offer an increase in coding gain.

Figure 5–3 shows the increase in gain as the constraint length increases for convolutional codes. This is for a rate of $r = 1/2$ with hard decision, that is, the decoding decision is made on a simple bit basis (see later section on soft decision). Note that in going from $K = 3$ to $K = 8$, there is an improvement in coding gain of about 1.6 dB at BER = 10^{-4}. Even though there is an increase in decoder complexity in going from $K = 3$ to

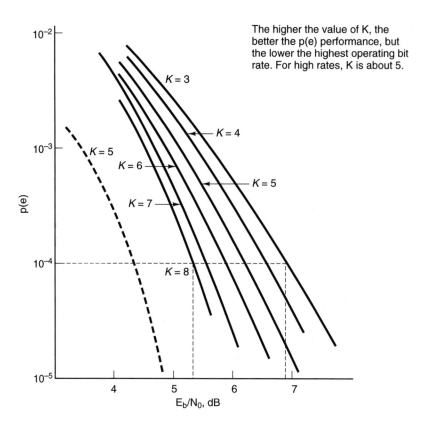

- r = 1/2
- Soft decision: Q = 3 (8-level quantization).

Figure 5–3 $p(e)$ vs. E_b/N_0 with parametric values of the constraint length, K. Shown for hard decision ($K = 3 \rightarrow 8$) and soft decision ($K = 5$).

$K = 8$, the gain of 1.6 dB coding can have a significant impact on system design and cost. Note that the 1.6-dB increase in gain is over and above that which results in going from no coding to the $K = 3$ design. The dotted curve is for 3-bit or eight-level quantization for $K = 5$. For $K = 5$ with hard and soft decision (several bits used in the decision process), the practical differential in gain is about 2 dB at BER = 10^{-5}. We can use a higher level of quantization, but the additional improvement is small.

If the demodulator before the decoder is designed to produce hard decisions (either 1 or 0), the subsequent channel decoding process is called *hard decision* decoding. This may also be referred to as binary quantization decision ($Q = 2$). Hard decision merely states that the transmitted symbol was 1 or 0 at the sampling interval, i.e., decision was made on every bit.

An alternative to hard decision is when the output of the demodulator is quantized to more than two levels. This decoding is designated as soft decision, or multi-level quantization. A decision is not made on each level. The demodulator first decides whether the output voltage is above or below threshold and then computes a confidence number that indicates the distance from threshold by quantizing the demodulator output. Several levels of quantization are possible, but usually eight-level or three-bit quantization is used and referred to as $Q = 3$ (2^3). This is the confidence of stating that a 1 or 0 has been transmitted. Soft decision achieves about a 2-dB practical improvement (in E_b / N_o) in coding gain over a hard decision at about BER = 10^{-5}. This gain is manifested on subsequent figures in this chapter.

It was shown earlier by the Shannon bound that the theoretical lower bound of E_b / N_o (for zero errors) is −1.6 dB. This can, in principle, be achieved if soft decision is used. If hard decision is used, the best that can possibly be achieved is +0.4 dB, or 2 dB higher than if soft decision prevails.

Figure 5–4 shows the results for a constant length and several code rate values for Viterbi hard and soft decision decoding for convolutional codes. In addition, sequential decoding with variable code rates and fixed constraint length is shown. Notice the improvement in performance with reduced rates, as well as using soft decision over hard decision. At BER = 10^{-5}, there is roughly a 2-dB improvement in going from hard decision to soft decision with the same rate and constraint.

You will notice that changing the code rate, r, of the convolutional encoder/Viterbi decoding does not change performance significantly. Only an increase in constraint length, K, will appreciably improve the coding gain. This latter benefit comes with an increase in decoder complexity, especially when negotiating the trellis when using the Viterbi decoding algorithm. More will be said on trellises in a later section.

Figure 5–5 is from a Linkabit FEC brochure. It shows the coding gain of convolutional encoder/Viterbi decoding for various code rates, r, as compared with uncoded PSK at BER = 10^{-5} with E_b / N_o = 9.6 dB. As indicated, the gains are markedly better than the other two less powerful block codes displayed.

The code $r = 1/2$ (100% redundancy) requires double the bandwidth of the uncoded system. If rate 3/4 is used, the redundancy is 33% and the bandwidth expansion is 4/3. The coding gain is about 1 dB less than for $r = 1/2$.

The curve at the bottom is the Shannon bound. Shannon's coding theorem is only an existence theorem: it does not give any clues concerning the coding scheme to use to reach this bound. That is left to the imaginative researcher. It is recalled that to get to the Shannon bound of E_b / N_o = −1.6 dB, for a white Gaussian noise channel, the PSK uncoded modulation requires a coding gain of 11.2 dB (9.6 + 1.6).

Additional performance curves for other coding schemes are represented in Figures 5–6 and 5–7. Both hard decision and soft decision are shown. Note for the codes

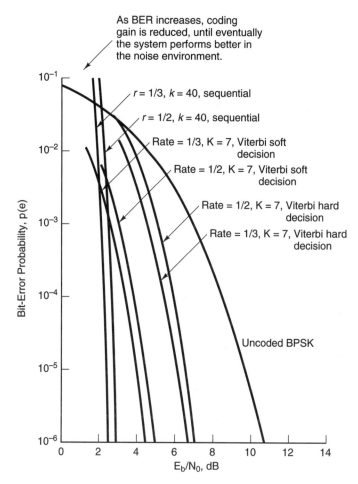

- Increase coding gain by:
  - Using lower rate (r) codes.
  - Increasing code complexity by increasing constraint length, K.
- Coding reduces required E_b/N_0 by
  - Increasing bandwidth if same modulation as the uncoded case is used; decreasing throughput if same bandwidth as the uncoded case is used.

Note: In the sequential decoder curve, the constraint length is large, but this is relatively independent of complexity.

Figure 5–4 Comparison of various decoding schemes for varying rates and constraint length, K.

that t is the number of errors corrected. For the Hamming codes, the error corrected is one. The symbolism (n, k) is defined previously, where $n - k$ equals redundancy bits.

Code Performance

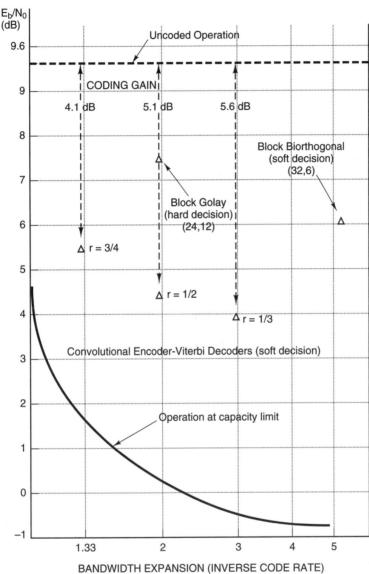

Comparative Performance of Decoders: Required E_b/N_0 for 10^{-5} Bit Error Rate.

Figure 5–5 Coding gains of convolutional encoding/vierbi decoding for various code rates, *r*, and comparison with several short block codes. Reprinted with the permission of Titan Linkabit.

Plots of several codes for error correction on random error channels are shown in Figure 5–8. Several codes of moderate complexity, showing moderate gain, are shown

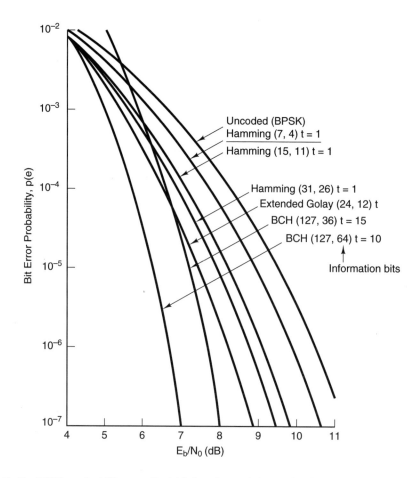

Figure 5–6 BER probability vs. E_b / N_o for demodulated BPSK over a Gaussian channel for several block code types.

on the right side of the plot. It is obvious that the Viterbi decoder offers better gain than the block codes shown. Longer BCH codes, however, can give gain performance similar to convolutional/Viterbi decoding (see Figure 5–7). The coding gains for BER = 10^{-5} for several codes are shown in Table 5–2. For a Viterbi decoder with $K = 7$, and $r = 1/2$, we notice a 2-dB improvement in using soft decision. The sequential decoder has a high constraint value, but implementation of the decoder is independent of K. It also has better performance than the Viterbi decoder for BER values less than 10^{-5}.

Note that the above cited codes work best in random noise channels. In a fading or multipath environment where many wireless systems operate, bursty-type errors are expected to occur. Many codes cannot cope with this type of perturbation. This can be

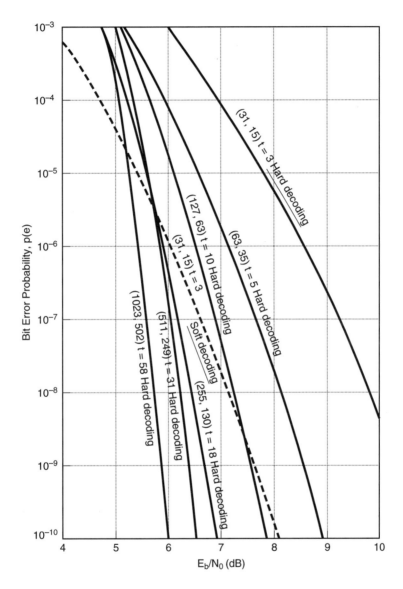

Figure 5–7 BER probability vs. E_b/N_o for demodulated BPSK over a Gaussian channel using Bose-Chaudhuri-Hocquenghem (BCH) codes.

combatted by the interleaving of bits at the transmitter and de-interleaving at the receiver. Also, we may use codes designed to cope with this type of impairment, namely Reed-Solomon (R-S) codes. This code type will be described in a subsequent section.

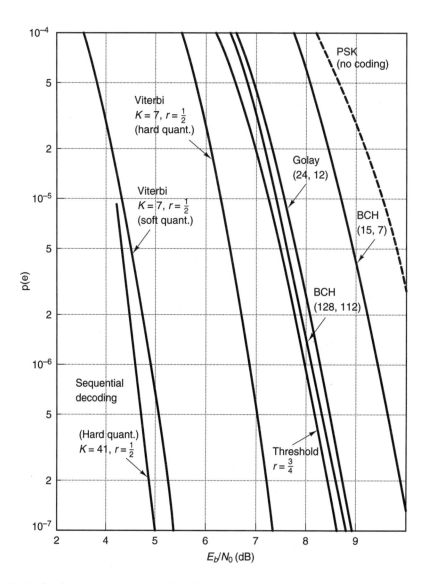

Figure 5–8 Performance of several coding schemes.

5.4 Block Coding

In block coding, as the name implies, a succession of data symbols are encoded independently into blocks of symbols for transmission. This is in contrast to convolutional encoders, which have no block structure. The received blocks are decoded independently into blocks of data symbols.

Table 5-2 Coding Gain for Several Coded Systems.

System	Required Value of E_b/N_o at $P_e = 10^{-5}$ (dB)	Coding Gain (dB)
Ideal PSK (no coding)	9.6	—
BCH (15, 7)	8.7	0.9
Golay (24, 12)	7.6	2.0
BCH (128, 112)	7.5	2.1
Threshold ($r = 3/4$)	7.4	2.2
Viterbi $K = 7$, $r = 1/2$		
Hard Quantization	6.5	3.1
Soft Quantization	4.5	5.1
Sequential $K = 41$, $r = 1/2$		
Hard Quantization	4.4	5.2

In block coding, an input data stream is read into the encoder in blocks of information bits containing k bits (see Figure 5–9). The output of the encoder contains the k bits, plus p check bits, giving an output code word of $n = k + p$ bits. This is referred to as an (n,k) encoder. The coding rate is given as the ratio of the information, k, over the n code word, or k/n. This is also the inverse of the bandwidth expansion factor. Clearly, the check bits are used in the receiver decoder to identify and possibly correct any errors. The more check bits, the better the BER achieved, but at the expense of increased bandwidth.

The correction of random error is dependent on the minimum number of positions in which any pair of encoded blocks differ. This has been referred to as the Hamming distance, d, of the code.[2]

For typical codes, the Hamming distances are as follows:

$$\text{Hamming}: d = 3$$
$$\text{BCH}: d = 2t + 1$$
$$\text{R-H}: d = 2t + 1 = n - k + 1 \tag{5.8}$$

where t is correctable random errors. Reed-Solomon codes will be discussed in a subsequent section.

---

2. Hamming distance: The distance between *two* code words is defined as the number of places in which the two code words differ. Not to be confused with Hamming weight, which is defined as the number of 1s in a code word.

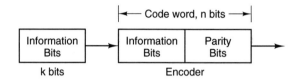

Where k: information bits.
n: block length.
(n–k): number of parity check bits.*
k/n: code rate; determines amount of redundancy;
also inverse of bandwidth expansion factor.

- A sequence of k information symbols are encoded in a block of n symbols, n > k.
- Block code is frequently denoted by (n, k, t), where t is the number of errors in a block of n symbols the code can correct.
- Each group of k consecutive information bits is encoded into a group of n symbols, which are transmitted. Each block of n symbols corresponds to a particular group of k information bits.

*In an (n,k) block code for code rate r, n-k redundant bits are added to each k-bit information sequence to form a code word. These n-k redundant bits allow the code to combat channel noise and are called parity bits.

Figure 5–9 Block code (*n*, *k*, *t*) generation.

Block codes are represented by the following parameters:

(n,k) : representation.

n : length of code word.

k : number of data bits (parity bits integrated with k).

$p = (n - k)$: number of parity bits.

$n = 2^p - 1$: length of bits in code word.

d : minimum Hamming distance of code word.

2^k : distinct code words.

$t = (d - 1) / 2$: number of correctable errors.

k/n : code rate; determines amount of redundancy.

The decoder processes individual blocks (memoryless). All require bandwidth expansion.

$$e.g., (n,k) = (7,4)$$
$$R_b \text{ info. bits} = 2400 \text{ bps (say)}$$
$$R_{\text{trans.}} = 2400 \times (7/4) = 4200 \text{ bps} \tag{5.9}$$

Therefore, an increase in bandwidth.

$$\text{Uncoded: } E_{bt} = PT_b = P / R_b = P / 2400 \tag{5.10}$$

$$\text{Coded: } R_{\text{info.}} = 2400 \text{ bps}$$
$$R_{\text{trans.}} = 2400 \times (7/4) = 4200 \text{ bps} \tag{5.11}$$

$$E_{b\text{trans.}} = P / R_t = P / 4200 \tag{5.12}$$

5.4.1 Popular Block Codes

Hamming code is one of the earliest and most popular error detection and correction codes [13]. It is not a powerful code, but it is able to detect and correct a single error. It can also detect two errors, but not *correct* them. The performance of this code in comparison to other more powerful block codes is shown in Figure 5–6.

Suppose we use a Hamming code to detect and correct errors. The code representation is (n,k), where n is the length of the code word and k is the number of information bits. Assume the number of data bits input to the encoder is 4, with the encoder integrating 3 parity bits. The code word length is therefore 7, which can also be found from $n = 2^p - 1$, where p is the number of parity bits. The code therefore assumes the identity (7,4). From this we can also ascertain the number of parity bits, $7 - 4 = 3$.

From the parameters on Hamming coding given previously, the number of code words which can be transmitted is $2^k = 16$. Since there are only 4 information digits, we can code only 16 symbols. The minimum Hamming coding distance, d, between code words is 3. The number of correctable errors is therefore $t = (d - 1) / 2 = 1$.

Now let us consider in more detail an example of the error correcting feature of the Hamming code using the parameters used above.

Assign positions x_3, x_5, x_6, x_7 as information bits in a code word. The remaining positions are to be occupied by parity bits. We have

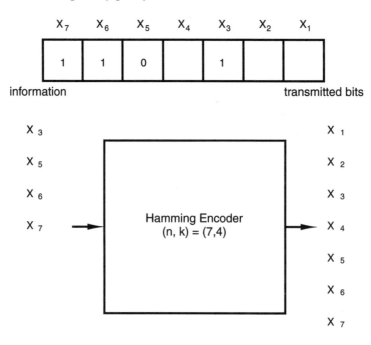

At the encoder we perform the following modulo-2 addition operation to fill in the parity bits (the sum must equal (= 0))³:

1. $x_1 \oplus x_3 \oplus x_5 \oplus x_7 = \text{even} = 0$

2. $x_2 \oplus x_3 \oplus x_6 \oplus x_7 = \text{even} = 0$

3. $x_4 \oplus x_5 \oplus x_6 \oplus x_7 = \text{even} = 0$

$$1 \oplus 1 = 0$$
$$0 \oplus 1 = 0$$
$$0 \oplus 0 = 0$$
$$1 \oplus 0 = 1$$

---

3. If all zeros, this equals an even number of 1s too.

Block Coding 87

Therefore, for this to be true (sums equal to zero), x_1, x_2 and x_4 must have the values indicated below:

1. $\underline{0} \oplus 1 \oplus 0 \oplus 1 = 0$
2. $\underline{1} \oplus 1 \oplus 1 \oplus 1 = 0$
3. $\underline{0} \oplus 0 \oplus 1 \oplus 1 = 0$

The transmitted code word is therefore where the three bits are given as $x_1 = 0$, $x_2 = 1$ and $x_4 = 0$.

X_7	X_6	X_5	X_4	X_3	X_2	X_1
1	1	0	0	1	1	0

Note that this is one of 16 code words with differing data bits and parity bits, and the above is performed for each of the remaining 15 code words. A listing of code words is shown in Figure 5–10.

1	2	3	4	5	6	7
0	0	0	0	0	0	0
1	1	0	1	0	0	1
0	1	0	1	0	1	0
1	0	0	0	0	1	1
1	0	0	1	1	0	0
0	1	0	0	1	0	1
1	1	0	0	1	1	0
0	0	0	1	1	1	1
1	1	1	0	0	0	0
0	0	1	1	0	0	1
1	0	1	1	0	1	0
0	1	1	0	0	1	1
0	1	1	1	1	0	0
1	0	1	0	1	0	1
0	0	1	0	1	1	0
1	1	1	1	1	1	1

- Positions 3, 5, 6, and 7 are information digits.
- Positions 1, 2, and 4 are parity check digits.
- The minimum Hamming distance is "3."

Figure 5–10 Hamming code words for $(n,k) = (7,4)$.

Now assume the detected signal is 1 $\underline{0}$ 0 0 1 1 0. We assume that position 6 (reading from the right) is in error. A "1" was transmitted, but the channel has caused an error and it is now a "0". Now the decoder must find this error position and correct it.

At the *decoder* in the receiver, we perform the same operation with the received incorrect code word. As before, the sum must be even (= 0) if the correct code word was received. If not, an error has occurred.

X_7	X_6	X_5	X_4	X_3	X_2	X_1
1	0	0	0	1	1	0

1. $x_1 \oplus x_3 \oplus x_5 \oplus x_7$
2. $x_2 \oplus x_3 \oplus x_6 \oplus x_7$
3. $x_4 \oplus x_5 \oplus x_6 \oplus x_7$

Using the received erroroneous code word, and performing Steps 1, 2, and 3, we obtain

1. $0 \oplus 1 \oplus 0 \oplus 1 = 0$
2. $1 \oplus 1 \oplus 0 \oplus 1 = 1$
3. $0 \oplus 0 \oplus 0 \oplus 1 = 1$

If the sum were all zeros, there would be no error. However, we noticed that the parity check is *not* satisfied. The error is in Position 6 since binary 110 represents the equivalent number in the decimal system as shown in Table 5–3.

Reed-Solomon (R-S) codes form a class of block codes. Unlike convolutional codes, the encoder operates on multi-bit *symbols* instead of individual *bits*. They are useful in correcting random errors, but are particularly powerful in combatting error bursts.

The input data is assembled in *n*-bit symbols, and *k* such symbols are encoded into *n* symbols with the symbol-correcting capability of

$$t = (n - k) / 2 \tag{5.13}$$

For example, for an R-S code $(n,k) = (63,47)$, the total number of transmitted symbols is 63, and the number of information symbols is 47. It can correct

$$t = (n - k) / 2 = (63 - 47) / 2 = 8 \text{ symbol of errors} \tag{5.14}$$

Consider the sketch shown in Figure 5–11, which depicts the encoding and decoding of 47 information symbols (each symbol contains 6 bits) using an R-S(63,47) coder. After R-S coder processing, the total number of symbols is 63. The total number of bits output is $6 \times 63 = 378$ bits per block.

The decoder in the receiver converts the 378 input bits to parallel form for processing by the R-S decoder. It then converts the bits to serial form to yield the original information in bits per block. Note that the decoder attempts to correct symbol errors rather than transmitted bit errors.

Table 5–3 Decimal Equivalents of Binary Numbers.

Binary Number	Equivalent in Decimal System
0	0
1	1
10	2
11	3
100	4
101	5
110	6 (e.g., $1 \times 2^2 + 1 \times 2^1 + 0 \times 2^0 = 6$)
111	7
1000	8
1001	9
1010	10
1111	15
10100	20

We indicated in the section on block codes that the correction of random errors is dependent on the Hamming distance manifested by encoded blocks. For the R-S code given here, the Hamming distance is given by

$$d = n - k + 1 = 63 - 47 + 1 = 17$$

This may also be determined from the number of correctable errors, t

$$d = 2t + 1 = 2(8) + 1 = 17$$

5.5 Convolutional Encoding

Convolutional codes[4] are named as such because the redundant bits are generated by modulo-2 convolutions. The encoder accepts an input sequence of bits and produces

---

4. Convolutional coding is used in cellular standards such as GSM, IS-54, and IS-95. It also has wide application in satellite networks.

Encoding (XMTR)

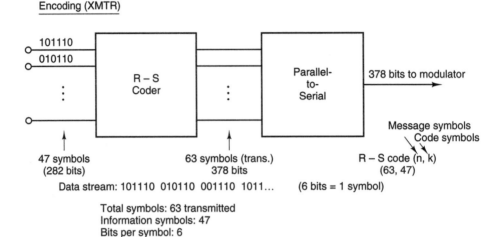

Data stream: 101110 010110 001110 1011... (6 bits = 1 symbol)

Total symbols: 63 transmitted
Information symbols: 47
Bits per symbol: 6
$6 \times 47 = 282$ information bits per block

Encoding expands the 47 symbols into 64 symbols.

$6 \times 63 = 378$ trans. bits per block

There are therefore 63-47 = 16 symbols of redundancy (96 bits).

Symbol-correcting capability:

$2t = (n-k) = (63-47) = 16$, $t = 8$ symbols

Can correct up to and including eight symbol errors in a block of 63 symbols.

Decoding (RCVR)

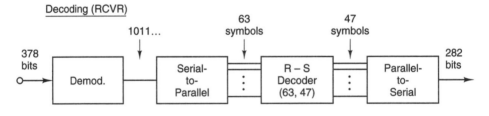

The Berlkamp-Massey algorithm is used to decode R–S codes.

Figure 5–11 Read-Solomon coding and decoding.

an output sequence with controlled redundancy to allow error correction. Each parity check is interleaved between each information bit. A block diagram of a rate $r - 1/2$ and $K = 3$ constraint length coder is shown in Figure 5–12.

The rate $r = 1/2$ indicates that a code bit is added for each data bit input. The smaller the rate, the greater the number of redundant bits added. For a rate $r = 1/2$, the

Convolutional Encoding

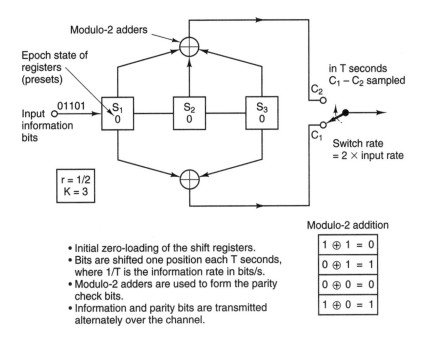

- Initial zero-loading of the shift registers.
- Bits are shifted one position each T seconds, where 1/T is the information rate in bits/s.
- Modulo-2 adders are used to form the parity check bits.
- Information and parity bits are transmitted alternately over the channel.

Modulo-2 addition

$1 \oplus 1$	$= 0$
$0 \oplus 1$	$= 1$
$0 \oplus 0$	$= 0$
$1 \oplus 0$	$= 1$

Figure 5–12 Convolutional encoder.

bandwidth required is twice that of the uncoded system design. For $r = 1/3$, there is a three-fold increase in bandwidth.

For three common coding rates, each data bit output is accompanied by additional bits as follows:

1/2 : for each data bit, there are 2 output bits.
1/3 : for every two data bits, there are 3 output bits.
3/4 : for every three data bits, there are 4 output bits.

Attendant with this FEC-type scheme is an increase in coding gain, which was shown in previous sections.

The constraint length, K, is given by the number of shift registers in the network, and the number of shifts (encoder output) over which a single information bit will influence the encoder output. The rationale for going to longer constraint lengths is that significant additional gain is obtained for all code rates. K of values up to about ten are feasible, giving larger coding gain (less E_b / N_o required to give same BER as an uncoded system). However, the processing becomes more complicated (especially in the decoder) with increases in delay. In the decoder, a large number of operations must be performed and the transmission speed has to be slow enough to allow propagation through the decoder. Greater throughput is possible with smaller values of K. More will be said on this in Chapter 6.

Convolutionally coded messages can be decoded with Viterbi decoders (using the Viterbi algorithm [1]), sequential decoders [11,12], or threshold decoders [5]. The most popular is the Viterbi decoder, which has been used extensively in satellite applications, and now in cellular systems. Viterbi decoding works best with random errors, but has difficulty coping with bursty-type errors. In wireless systems, errors of this type frequently occur in fading environments. However, there are schemes which can combat bursty errors, such as interleaving (breaking up clusters of errors) or the use of R-S codes. The latter is frequently concatenated with either a convolutional encoder or a block code. Therefore, these schemes complement each other with R-S principally taking care of the bursty errors and the inner code coping with the random errors. Frequently, interleaving is also used with R-S coding. A typical configuration for this concatenation was shown previously in Figure 5–2.

At the output of the encoder in Figure 5–12, a commutator is used to alternately sample the output of the two modulo-2 adders. For example, the output of the upper modulo-2 adder is $C_2 = S_1 \oplus S_2 \oplus S_3$. The lower gives $C_1 = S_1 \oplus S_3$. Therefore, the first data bit yields coded output bits, $C_1 + C_2$.

In Figure 5–12, the information bits are shifted to the right though the cascaded shift registers [3] as the next information bit proceeds from the left. For the configuration shown, $n = 2$ output symbols are generated for each input bit yielding a code rate, $r = 1/2$.

The operation of the convolutional encoder is fairly straightforward. Consider the encoder in Figure 5–13, which has an input data stream of 0 1 1 0 1. Note that since $r = 1/2$, each input data bit will yield two output bits, assuming initial loading of the registers is all zeros. This is not a firm requirement, however, since any loading may be used, but this simplifies the explanation.

The first bit into the structure is "0". Each time a new bit is shifted into the encoder, the modulo-2 adders' outputs are sampled by the multiplexer. The bits in the register are shifted one position to the right each T seconds, where $1 / T$ is the information rate in bits per second.

For example, in Figure 5–13, for the first bit into the structure, which is "0", the number one position of the multiplexer yields the modulo-2 addition from the S_1 register dump with the S_3 register dump. We therefore have

$$C_1 = 0 \oplus 0 = 0$$

For position two of the multiplexer, we modulo-2 the adder dump from S_1, S_2, and S_3. Since the register was loaded with all zeros at the start, and the input was zero, the addition becomes

Convolutional Encoding

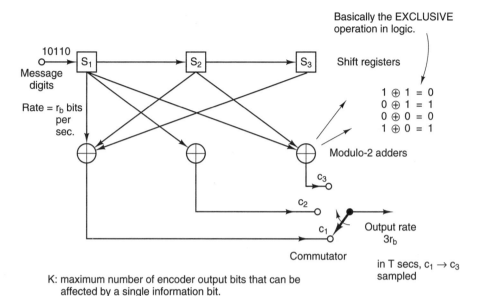

K: maximum number of encoder output bits that can be affected by a single information bit.

- Modulo-2 adders are used to form the parity check bits.
- Each time a new message bit is shifted into the encoder, the mod-2 adders' outputs are sampled by the commutator.
- Constraint length of code, K = 3 (in this case).
- Three output bits are generated for each input bit yielding a code rate of r = 1/3.
- The bits are shifted one position each T seconds, where 1/T is the information rate in bits per second.
- For example, assume the message is indicated in binary bit form indicated at input to the shift register bank above:

For the initial message, bit "1" is loaded* into the encoder, and the commutator terminal outputs yield:

$c_1 = S_1 \oplus S_2 \oplus S_3 = 1 \oplus 0 \oplus 0 = 1$
$c_2 = S_1 = 1$
$c_3 = S_1 \oplus S_2 = 1 \oplus 0 = 1$

Therefore, output = 111 for input message bit "1".

Message Bit	Out
1	11
0	10
1	01
1	01
0	00

* Initial loading of registers is zero (even though it does not have to be).

- Coding increases power efficiency
  - But increases bandwidth if the same modulation as uncoded case is used.
  - But decreases throughput if same bandwidth as uncoded case is used.
- Coding gain increases by
  - Using lower rate codes (code rate above is 1/3); not significant improvement.
  - Increasing code complexity by increasing K (more shift registers). The present limit of Viterbi decoders is K ≈ 10. However, for space applications, K as high as 31 has been used.

Figure 5–13 Example of a convolutional encoder.

$$C_2 = 0 \oplus 0 \oplus 0 = 0$$

Therefore, for an initial "0" input of the data stream, we obtain an output of 0 0.

The next data bit input from above is "1". After negotiating the encoder, this yields 1 1. If we continue along in this vein for the rest of the input data bits, we obtain at the output

$$0\ 0\ 1\ 1\ 0\ 1\ 0\ 1\ 0\ 0$$

resulting from 0 1 1 0 1 input data bits. Note that there are twice as many output bits as input bits, as we expected for a rate of, $r = 1/2$.

Figure 5–13 is also an example of a convolutional encoder using three shift registers. Note here the rate is $r = 1/3$ and the output therefore contains two additional bits besides the data bit. For five-bit data input, we would obtain fifteen output bits. The redundant bits are used by the decoder to recover the original data.

Another method of representing the encoder output sequence is via a coding tree, which is shown in Figure 5–14. Assume the input data bits are 1 0 1 1 1. At each node of the tree, the information bit determines which direction it will take — for example, down for a "1" or up for a "0". For the first input data bit of "1", the transition path is to path "C" and the output is 1 1 (from the encoder). The second data bit goes up the tree from position "C" to "F". Note that the third data bit is "1", so the route is down to "I". The remaining two bits in the data word go down in the tree. The n (from $r = 1/n = 1/2$) digits occurring in the branch selected correspond to the output symbols of the encoder. These are indicated by the ascended path, 1 1 0 1 0 0 1 0 0 1, for the example shown.

A further help in visualizing the process of decoding is via the use of a trellis, as shown in Figure 5–15. The trellis reveals all routes through the code tree. For the example shown, there are eight possible paths. From the sequence, *we chose the path which has the least Hamming distance from the received sequence*.

A method used to decode the sequence is the Viterbi algorithm. This reduces the number of computations required. The Viterbi algorithm uses a maximum likelihood decoding technique, that is, the incoming code stream into the decoder is compared against every possible path through the trellis and the one which is closest to it defines the output data. Again, finding the shortest path through the trellis is the goal. This path is the Hamming distance[5].

As an example, consider the following: A maximum-likelihood decoding scheme examines a received sequence, *y*, and selects from a stored set of possible code words

---

5. A minimum Hamming distance decoder is a decoder that finds the code word that differs from the received word in the fewest places. The Hamming distance between code words is the number of positions in which they differ. For example, for 1 1 0 1 0 1 and 1 1 1 0 0 0, the Hamming distance is 3. The Viterbi algorithm minimizes the Hamming distance.

Convolutional Encoding

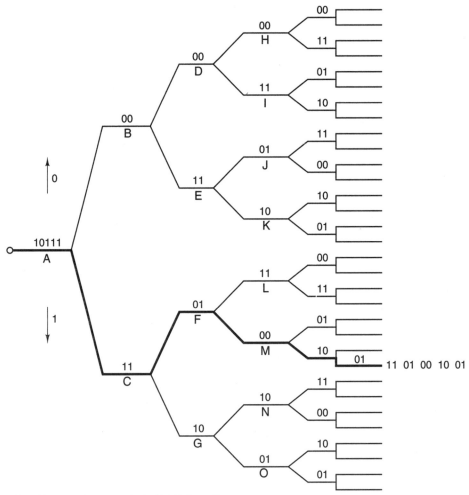

- Data bitstream into encoder is 10111 (note first entry into tree is down due to initial "1" of data word).
- Sequence output of the encoder is 11 01 00 10 01.
- Double digits refer to the encoder output (because r = 1/2).

Figure 5-14 Coding tree for binary convolutional encoder.

that code word x has the minimum distance from y. The maximum-likelihood decoder is therefore a "minimum distance" (Hamming) decoder.

For example, consider the following bit sequences to be a set of possible code words:

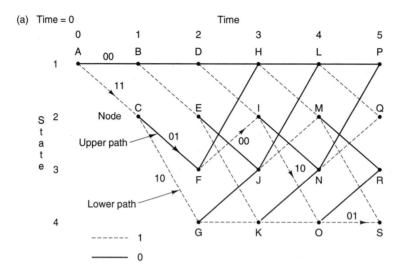

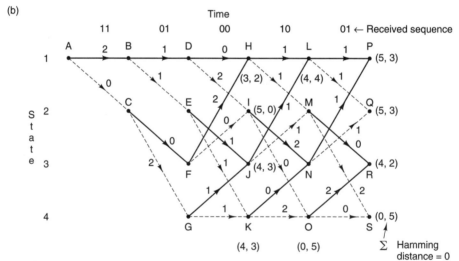

- Superimposed Hamming distances.
- Discard paths having higher Hamming distances.
- Note the path ACFIOS (Σ Hamming distance = 0).
- A message which is a sequence of symbols can be described as a particular path through the trellis.
- Each dot represents a node in the trellis.
- Successive notes are separated in time by one signaling interval.
- Note from each node that there are two possible transitions, or two possible symbols.

Figure 5–15 Trellis representation of the code tree shown in Figure 5–14.

Convolutional Encoding

$$x_1 = 0\ 0\ 0\ 0\ 0$$
$$x_2 = 1\ 0\ 1\ 0\ 1$$
$$x_3 = 0\ 1\ 0\ 1\ 1$$
$$x_4 = 1\ 1\ 1\ 1\ 0$$

Assume the received sequence is $y = 1\ 0\ 0\ 0\ 1$. The maximum-likelihood (minimum-distance) decoder will select y as x_2 since the distance between y and x_2 is 1, while the distance between y and any other transmitted sequence is greater than 1.

In minimum-distance decoding, the decoder stores a "table," or code book, containing all the allowable binary sequences of the specified code word. Each of the received sequences is assigned to the stored code word that most closely matches it.

The trellis in the lower half of Figure 5–15 was redrawn to show the Hamming distance for all possible paths through the trellis, that is, comparing the received bits displayed at the top of the trellis with those in each branch. For example, path AC, which is labeled "1 1" in the top trellis, gives a Hamming distance equal to "0" on the bottom trellis. This is done for each path and the Hamming distances summed.

Notice that some of the junction numbers are indicated in parentheses. These are cumulated Hamming distances from different paths arriving at that junction. For example, take the point (5,0) shown in Figure 5–15(b). The five results from the cumulated distance from the upper path ABDI (i.e., 2 + 1 + 2 = 5). The second number in parentheses results from the path ACFI (0 + 0 + 0 = 0).

Since we are interested only in the paths with the minimum distances, we can at this point discard the paths with the larger distances. If a junction has equal distances, either one can be discarded. Rejecting paths and retaining the survivor path is part of the strategy in the Viterbi algorithm. From Figure 5–15(b), the minimum distance of cumulative zero is given by the path ACFIOS. This results from the data input to the encoder of 1 0 1 1 1.

In convolutional (and block) encoding, redundant symbols are added to source symbols, and decoding is based on minimizing the Hamming distance. It may be interesting to note at this junction that coding is done *independently* of the modulation process. However, in some seminal work done by Ungerboeck [14], he has shown that it is more efficient to integrate modulation and coding, as in a technique known as trellis coded modulation (TCM). In his elegant approach, convolutional encoding/Viterbi decoding and modulation can be combined without a sacrifice in bandwidth. However, we will not discuss this concept here.

In summary, three decoding algorithms (with varying degrees of power) are frequently used with convolutional encoders. They are:

- Threshold decoding (for threshold-decodable codes).
- Viterbi decoding (for short constraint-length codes) long constraint-length codes make decoding very complicated; the trellis is extensive.

- Sequential decoding (high constraint length compatible with high coding gains).

5.6 Concatenated Coding

Two or more codes can be combined to form a more powerful code. For a concatenation of two codes, the codes are referred to as inner and outer codes.

In a mobile environment, several kinds of errors may occur due to fading and multipath, including both random error events and outages due to cluster errors. In concatenation, two codes are designed to complement each other where one code combats random errors and the other is more powerful in coping and correcting bursty-type errors.

Two codes that are frequently used in this combination are Reed-Solomon (R-S) codes (block code) and convolutional codes (C-C). Convolutional codes are used to reduce random errors and are ineffectual against bursty errors, while the R-S code is powerful in reducing burst-type errors. The former code is usually the inner code. Interleaving may also be imposed to further assist in removing bursty errors via bit randomization. The interleaving may be separate from the two codes, but may also be incorporated into the outer coder (R-S). The location of the interleaver is shown in Figure 5-15(b).

In the decoding process, the Viterbi algorithm decoder is used to remove random errors. The sequential decoder may also be used for this application. The decoder used for the R-S code is the Berelkamp-Massey algorithm.

Figures 5–16 and 5–17 illustrate the dramatic performance improvement of a concatenated R-S and C-C as compared with a convolutional encoder alone. Roughly 2-dB improvement at 10^{-5} is indicated. Notice that the concatenated system has excellent BER performance with concomitant low bit error rates, whereas convolutional codes perform best with BER in the 10^{-5} range. Better BERs are possible, but the constraint length becomes high and decoding becomes extremely difficult.

The performance curve shown in Figure 5–17 results from a system used in NASA's deep space communications. Some parameters for the two codes, R-S and C-C, are indicated on the curve. Notice that for BER = 10^{-6}, there is an 8-dB coding gain from the uncoded PSK within 4 dB from the Shannon bound. There is a lot of redundancy here, but when you are communicating billions of miles, any savings in RF power or operation at very low E_b / N_o (via coding) is critical.

5.7 Interleaving

Another popular scheme to combat bursty errors, occurring in cellular communications, which may result from several channel impairments, is to incorporate interleav-

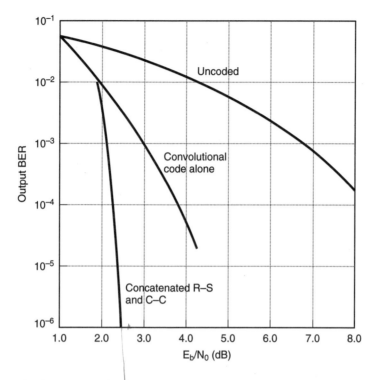

Figure 5-16 Performance of RS(255,223) and $(n,k) = (2,1)$, $K = 7$ convolutional code.

ing. This is used with convolutional/Viterbi decoders since they cannot cope with bursty errors. It is also used with R-S coding, even though bursty error correction is its forte, especially where bursts exceed the R-S error correction capability, or to overcome some anomalous behavior of the Viterbi decoder, if used in concatenation. Actually, interleaving complements R-S coding.

Interleaving randomizes (permutes) encoder output according to a fixed algorithm so that the bursty errors picked up by the receiver are more or less distributed over the entire coded waveform, and so that the lost data is not in a continuous chain of errors.[6] At the receiver, after demodulation, inverse permutation at the interleaver is performed before decoding. Thus at the decoder input, the data is not in the same sequence as the corresponding encoder output, but randomly spread over the frame. The decoder there-

---

6. A rather interesting and subtle comment was made by Bhargava [5] and Berlekamp [8] regarding interleaving. That is. interleaving is an information-destroying process which increases the entropy of the noise. By randomization, we are increasing the noise-like properties of the code. In the information-theoretic sense, random noise is the worst kind of noise. Structured noise (bursts) is easier to deal with than random noise. Noise is something we are combatting. So why make more of it? A rather profound thought to be entertained by the reader.

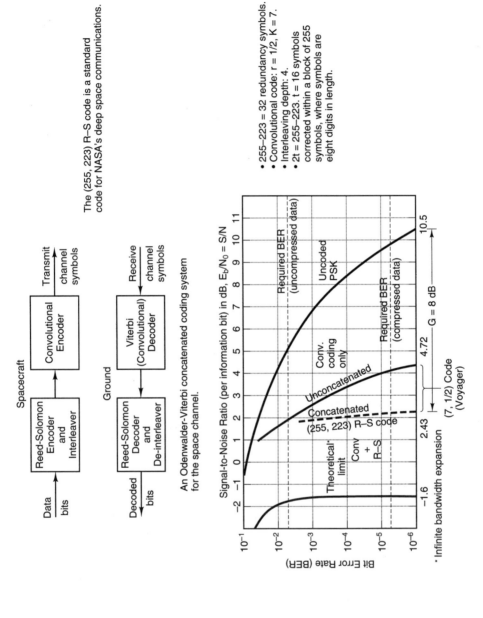

Figure 5–17 Typical performance curves for concatenated and unconcatenated coding systems for the space channel.

Interleaving

fore processes these scattered errors as random events. Note that interleaving must occur before the waveform is delivered to the decoder.

An example of interleaving in a block coding system is indicated in Figure 5–18.[7] As shown, a code word of n symbols is deployed as rows A, B, C, ..., N in a matrix form. The codes are w wide and the depth (A, B, ...) is a deployment of d rows. The code words consist of information bits and parity bits (n,k).

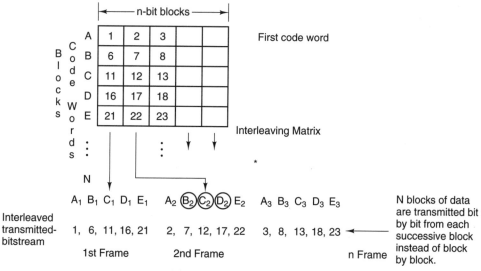

Figure 5–18 An example of block interleaving.

The output of the coder is now read out bit by bit and transmitted one *column* at a time. An example of this operation is shown in Figure 5–18. For the code in the matrix, the interleaved bitstream output is given as:

A_1 B_1 C_1 D_1 E_1 ... A_2 B_2 C_2 D_2 E_2 ... A_3 B_3 C_3 ...

1 6 11 16 21   2 ⑦ ⑫ ⑰ 22   3 8 13 18 27

---

7. This is the interleaving scheme used in the European GSM cellular standard.

The first frame is given as $A_1 B_1 C_1 D_1 E_1 \ldots$. For the example given here, if the blocks end at E, the interleaving depth is 5.

For ideal interleaving, we want columns and rows to be very big, but this introduces a lot of delay, which for certain applications may not be acceptable.

Now, suppose bursty errors have occurred in the transmitted bitstream cluster $B_2 C_2 D_2$ circled in the train above. At the decoder in the receiver, these will show up as *isolated* errors in the code blocks B, C, and D. We therefore have randomized the string of errors and they will appear at the receiver as noise-like events with which the decoders can cope. If the error burst is greater than d depth, say by a factor of j, then there will be no more than j errors in the code word. Thus, the greater the interleaving depth (5 here), the longer the burst that can be corrected, and the greater the redundancy and delay.

There are other block interleavers which operate more or less on the same principle, but with varying degrees of delay and complexity. The more complex the system, the greater the delay.

To summarize,

- Interleave the data and redundant bits after encoding.
- Perform the inverse operation at the receiver prior to decoding.
- Errors appear to the decoder as random events as opposed to a cluster of errors.
- The number of cluster bits corrected depends on the coding depth.
- The greater the depth, the greater the delay.

5.8 Coding Break-through

Recent developments in coding theory have presented a new class of convolutional codes called Turbo codes [16]. Results indicate that performance in terms of BER are close to the Shannon bound. Simulations suggest to within 0.7 dB of the Shannon limit can be achieved. Further elaboration is not within the scope of this chapter.

5.9 Conclusions

Telecommunications is afflicted with a paucity of spectrum, but also a shortfall in power in some applications. In particular, the latter is witnessed in satellite applications, where power in the space segment is difficult to come by.

To alleviate the spectrum scarcity, more efficient modulation is now being employed and researched, which will provide greater bandwidth efficiency in terms of bits per Hz. In addition, greater spectral efficiency is realized by producing waveforms in which spectra are mostly confined to the operational bandwidth (low and even nonexistent spectral sidelobes). Less inter-symbol and inter-channel interference results.

Some of these concepts have been reported in technical literature and should be referred to for additional details.

Interestingly, these higher order modulation waveforms offer greater capacity and migrate closer to the Shannon bound. However, to move increasingly closer to the bound, there is much more research to do in this area, which also *requires assistance from coding*. At Shannon's bottom line, he indicated that *efficient coding is necessary*. However, in his seminal writings, he did not indicate what coding should be used. Since his epochal paper in 1947, information theorists have been struggling to get closer to this theoretical bound. A review of applicable literature indicates that considerable progress has been made, especially some of the work done by NASA in this area.

Some of the newer modulation waveforms provide greater bandwidth efficiency (b/s-Hz), but also less communication efficiency, that is, greater power is required to achieve the desired BER. In addition, a signal state space diagram indicates that signal points in the constellation are closer together, giving rise to possible ISI due to denser decision space boundaries. Clearly this requires highly linear channels and/or compensating networks.

Forward error correction, which we discussed earlier, unfortunately vitiates some of the hard-earned gains made in improving bandwidth efficiency. The price paid by using FEC is bandwidth expansion because of the overhead due to redundancy required to correct the errors. Clearly there are opposing forces at play here. However, where there is adequate or barely marginal spectrum, a small dB coding gain helps to ameliorate the loss of some spectrum. There is credence to this comment since many of the wireless communication systems coming to fruition today and in the future are using or proposing to use FEC.

On the other hand, for those die-hards who want to hold on to their spectrum very tenaciously, there is now trellis coded modulation (TCM), which is finding increased use. TCM *maintains* hard-earned spectrum via that obtained by modulation research and by achieving coding gain at the same time, *with no additional spectrum required*. TCM is truly a breakthrough in communications.

In the implementation of forward error correction, it is noted that overhead is required to implement the code. The implication is that additional bandwidth is required. In a power limited, but not band-limited channel, this additional bandwidth is available. However, where a system is band limited, it may be difficult to implement coding. A seminal advancement in coding theory has been made entitled trellis coded modulation in which additional band-width in not required to realize coding gain.

As will be shown, adding symbol states in the state space (constellation) to increase the spectral efficiency, *integrated* with convolutional encoding, will achieve coding gains. This gain is in excess of the increase in E_b / N_o required to compensate for the loss in BER performance due to additional points in the constellation (points are closer together). This is the subject of Chapter 6.

5.10 References

[1] A. J. Viterbi, "Error Bounds for Convolutional Codes and an Asymptotically Optimum Decoding Algorithm," *IEEE Trans. Information Theory*, April 1967.

[2] J. A. Heller, and I. M. Jacobs, "Viterbi Decoding for Satellite and Space Communications," *IEEE Trans. Communication Technology*, October 1971.

[3] A. J. Viterbi, "Convolutional Codes and Their Performance in Communication Systems," *IEEE Trans. Comm. Technology*, October 1971.

[4] Special Issue on Coded Modulation, *IEEE Communication Magazine*, December 1991.

[5] V. K. Bhargava, et al., "Digital Communications by Satellite," Wiley-Interscience Pub., New York, 1981.

[6] Special Issue on Coding, *IEEE Communication Technology*, Part 2, October 1971.

[7] J. F. Hayes, "The Viterbi Algorithm Applied to Digital Data Transmission," *IEEE Communication Magazine*, March 1975.

[8] E. R. Berlekamp, et al., "The Application of Error Control to Communications," *IEEE Communications Magazine*, April 1987.

[9] V. K. Bhargava, "FEC Schemes for Digital Communications," *IEEE Communication Magazine*, January 1983.

[10] Linkabit Corp., "Error Control Products Catalogue."

[11] J. M. Wozencraft, "Sequential Decoding for Reliable Communication," Doctorate dissertation, MIT RLE, Report 325, 1957.

[12] J. M. Wozencraft, "Sequential Decoding for Reliable Communication," *1957 IRE National Convention Record*, Part 2, New York, 1957.

[13] R. W. Hamming, "Error Detecting and Error Correcting Codes," *Bell System Technical Journal*, April 1950.

[14] G. Ungerboeck, "Channel Coding and Multilevel/Phase Signals," *IEEE Trans. Information Theory*, January 1982.

[15] B. Pattan, "Information Theory School Notes," 1965.

[16] C. Berrou et al., "Near Shannon Limit Error-Correcting Coding and Decoding: Turbo-Codes," ICC May 1993, Geneva, Switzerland.

CHAPTER 6

Trellis Coded Modulation (Codulation)

6.1 Introduction

Radio spectrum is a limited natural resource. In recent years, interest has grown in the area of bandwidth-efficient modulation methods. These new modulation methods have made it possible to more closely approach the Shannon capacity limit. In previous chapters, we discussed how to increase the bandwidth (information density) and spectral efficiencies. We further explored methods to increase the communication (RF power) efficiency via coding, where coding gain is defined as the reduction in required E_b / N_o for a given error probability.

In satellite applications and mobile terrestrial systems, power is still a limited commodity. However, as is well-known in classical coding techniques, in particular in FEC, extra bits must be added (parity bits) to the transmitted symbol sequence, with the modulation operating at a higher data rate and hence requiring a larger bandwidth. For example, for a convolutional encoder of rate 2/3, the bandwidth increases by a factor of 3/2, i.e., by an amount equal to the reciprocal of the code rate. The bandwidth efficiency can be improved by the use of Pulse Amplitude Modulation (PAM), M-ary PSK, and QAM. Here, the information rate per transmission bandwidth is more than 1 bit/s-Hz. But, note that these higher order modulation methods result in signals which are closer together in signal space and require higher power levels to maintain the same detectability performance.

In the 1970s, telephone modems at 9.6 kbps were considered to have realized practical data rate limits. This was before the advent of Trellis Coded Modulation

(TCM) [1]. This technique has been successively applied to voice band modems, where it allowed the transmission of higher data rates than seemed possible before.

In 1982, a breakthrough emerged [1] which combined channel coding and multi-level/phase modulation to obtain a signal design that achieved significant power gain (3 to 6 dB) without any bandwidth expansion. This breakthrough led to TCM, which provides coding gain with no increase in bandwidth or reduction in data rate.

Today, through the use of TCM, modems at 56 kbps are now available. The first use of TCM for voice-grade commercial modems was found in the Codex 2360 and 2660 modems introduced in 1984. TCM has also found its way into international standards and satellite systems planning. Why satellites? Because in digital satellite communications, information rate requirements grew so fast that the bandwidths available manifested bandwidth limitations, as opposed to earlier designs where power was limited. In a more recent application, the U.S. digital television standard for HDTV transmission uses TCM combined with digital vestigial sideband modulation.

It is true that in a bandwidth-limited environment, bandwidth efficiency can be achieved by using higher order modulation methods (e.g., 8-PSK in lieu of 4-PSK), but in these higher alphabet signals, more power is required to maintain the same BER (the points are closer in signal space). Can we achieve the prior BER performance level before reverting to the higher order modulation? Yes, by coding, but we negate the bandwidth benefit we are seeking. Classical coding does not give us the results we want; therefore, TCM is used to obtain the advantages of error correction without incurring the penalty of increased bandwidth.

In classical coding schemes (convolutional or block codes), coding and modulation are treated as separate entities. Coding is performed on the original binary data stream independent of the modulation that follows. Here, coding is aimed at increasing the Hamming distance [19] between code words by adding redundancy to the initial information bits. There is an additional bandwidth required over and above that necessary for the information bandwidth, since extra parity bits must be added to the transmitted symbol sequence. As alluded to previously, there is a sacrifice in bandwidth, but an increase in power efficiency.

In TCM, coding and modulation are not separate and are used to maximize the minimum free Euclidean distance between transmitted symbol sequences, not the Hamming distance as in classical coding techniques. This concept will be elucidated upon later.

Since we are now considering bandwidth-limited systems, the redundancy required for coding is obtained by increasing the alphabet size, that is, the number of code symbols.

From Ungerboeck's classic paper (from which the curve in Figure 6–1 was taken), it can be seen that if the alphabet size is increased, for example, from 4-PSK to 8-PSK, and if the same information rate (bandwidth) is maintained, we realize an improvement in RF power efficiency. For example, for the *same* information rate of 2 bits/T seconds, going from 4-PSK to 8-PSK we achieve roughly a 6-dB reduction in signal-to-noise (S/N) ratio (note this is not E_b/N_o, which differs from S/N by 3 dB). By doubling the number of available signals (using 8-PSK), two bits per symbol may be conveyed with a $S/N = 6$ dB, provided some form of coding is used. Therefore, the redundancy required to realize this coding gain is achieved by increasing the number of coded symbols instead of the bandwidth. Doubling the number of symbols in the signal constellation by using TCM, two bits per symbol (4-PSK) can be transmitted while maintaining the same symbol rate. Since there is no change in the symbol rate, the coded system will have the same bandwidth as the uncoded system.

Note that if we were to increase the signal constellation in the absence of coding, the points in the signal space constellation would be closer together and would degrade performance because the error rate would increase (assumes the same power). However, the advent of TCM permits the improvement of the previous performance with no sacrifice in bandwidth, data rate, or power, because the coded signals actually move farther apart from each other.

As will be shown below, coding is optimized so as to maximize the minimum *free Euclidean distance* (ED)[1] between transmitted symbol sequences. Because coding gain is determined by the free Euclidean distance, the minimum ED of the original signal space is no longer a measure of error performance. Coding gain will occur whenever the free ED of the *coded* system is greater than the minimum ED of the reference system (pre-coded signal set). Ungerboeck's pioneer work shows how this is done.

6.2 The Theory

An example of a generic encoder/modulator for the TCM method is shown in Figure 6–2(a). The input is uncoded data bits with two symbol points in a signal space plane diagram. The convolutional encoder is also referred to as a trellis coder. Block encoding may also be used for this application, but convolutional codes (trellis codes) prevail since they allow for greater coding gain. Several of the bits are encoded and the output of the encoder is several uncoded bits plus bits which have been encoded. The extra bits (one or more) are parity bits, that is, the size of the modulation signal set is

---

1. Defined as the minimum Euclidean distance between any pair of transmitted code sequences which split at some level (in the trellis) and merge later.

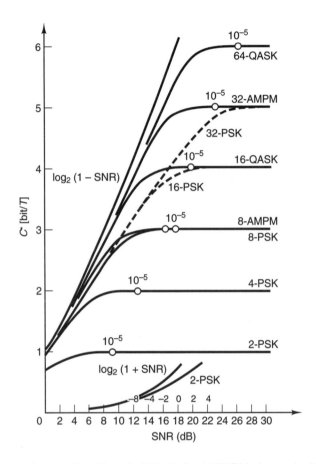

Figure 6–1 Channel capacity of bandwidth-limited AWGN channel with discrete value input and continuous-valued output. Copyright © 1982, 1989 by the Institute of Electrical and Electronics Engineers, Inc. Reprinted with permission.

increased from 2^n to 2^{n+k}, where n is the number of information bits and k is the number of parity bits introduced during each modulation interval. Ungerboeck [1] showed that by doubling the number of modulation signals (from 2^n to $2^{n+k} = 2^{n+1}$) one can achieve almost all the capacity that is possible. In other words, increasing the signal space constellation beyond doubling does not buy much more gain.

The $n + 1$ code symbols are therefore mapped into a set of two coded waveforms. As indicated previously, this expands the signal space constellation. For example, for the uncoded case of $2^n = 2^6$ symbols (64 QAM), the constellation goes to $2^{n+1} = 2^{6+1} = 128$ symbols for the coded case. This will provide significant coding gain without bandwidth expansion when compared to uncoded, conventional two-point sig-

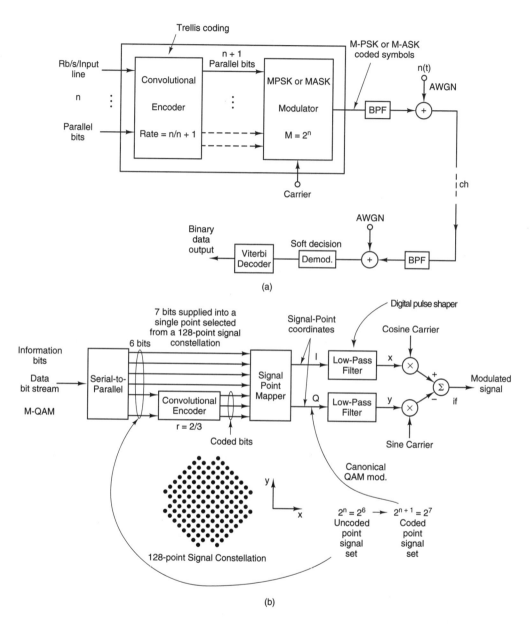

Figure 6–2 (a) Generic encoder/modulator/demodulator/decoder for TCM. (b) 14.4 kbps TCM transmitter (CODEX).

nal sets. Some form of coding must be used. Since the bandwidth is restricted, the redundancy for coding must be obtained by increasing the alphabet size. Figure 6–1 shows the reduced power required (at a fixed bits/T-sec value) in going from 4-PSK to 8-PSK. Gain can be achieved by doubling the signal alphabet and using appropriate code.

Even though an M-ary QAM signal was used as an example above, this also applies to M-ary PSK or PAM signals. Therefore, for an uncoded 4-PSK (QPSK) signal set with one parity bit (does not have to be one), where $n = 2$, the signal set is expanded from $2^n = 2^2$ to $2^{n+1} = 2^3$, or an 8-PSK constellation. The code rate in the encoder is $n / (n + 1) = 2/3$ (two data bits encoded to three bits, with one parity bit). There is no change in symbol rate. The coded system will have the same bandwidth as the uncoded system and will carry the same number of information bits per symbol (n), that is, $n = 2$ bits/symbol. In both systems, the coded 128 QAM and coded 8-PSK signals above, the coding gain will be expressed relative to the E_b / N_o requirements for the original uncoded 64 QAM and 4-PSK, respectively.

The following observations should be made. The signal set expansion (alphabet size increase) provides redundancy for coding. In the TCM scheme, coding and signal mapping functions are designed *jointly*, not separately as in conventional convolutional/block coding.

A more practical functional block diagram is shown in Figure 6–2(b). The TCM scheme adds a redundant code bit to the user data in each symbol interval as the data enters the encoder. At 14,400 kbps, the transmitter collects user data bits into 6-bit symbols at 2400 times per second, and encodes two of the six bits.

Notice that one redundant code bit is added to the user data symbol interval as this data enters the encoder. Two of the bits shown are convolutionally encoded in a rate 2/3 code so that seven total bits are derived from the six data bit symbols at 2400 times per second. That is, the encoder adds a code bit to the two input bits to form three encoded bits in each symbol interval. The three encoded bits and four data bits (uncoded) are mapped into a signal point selected from a 128-point [2] signal constellation. This is further detailed in a later section.

The constellation for the *coded* signal is shown in Figure 6–2, which is a 128-point signal constellation. This signal is processed by alpha filters and then modulated for transmission. At the receiver, the signal is demodulated and drives a soft decision Viterbi decoder (if the encoder is a convolutional encoder). The data is output with coding gain. Convolutional decoding using soft decision with three bits of quantization performs approximately 2 dB better than using hard decision.

You may have made the observation that the EDs (separation between points in signal space) of the coded constellation are closer than they are for the uncoded 64-symbol constellation. At prima facie, you would think that overall performance would be reduced because the error rate of the 128-point constellation is greater than the 64-point constellation (proximity of the signal points). However, a good code will "override" this reduced performance and provide coding gain. Therefore, *the EDs in signal space between individual symbols do not govern performance of the signal, but instead its performance is dependent on the minimum distance between the allowable sequences of symbols in the trellis*. This is the so-called "free Euclidean distance" of the trellis. This has been referred to previously and more will be said about it later.

Ungerboeck established signal set partitioning rules in which coding gain can be assured by assigning trellis transition branches to waveforms. That is, the free Euclidean distance between transmitted signal sequences will always exceed the minimum distance of the *uncoded* reference waveform.

Partitioning of a modulation signal constellation into subsets having increasing minimum EDs $d_0 < d_1 < d_2 < ...$, between elements (points in signal space) is required. These distances, coupled with the trellis, result in net coding gain. This is referred to as asymptotic coding gain (ACG) and is completely determined by the minimum free Euclidean distance, d, as follows:

$$\text{ACG} = 20 \log_{10} (d_{\text{free}} / P_{\text{ave,coded}}) / (d_{\text{ref}} / P_{\text{ave,uncoded}}) \qquad (6.1)$$

where d_{free} is the free Euclidean distance (of coded waveform), i.e., the minimum distance among all sequence paths that diverge and reemerge in the trellis, d_{ref} is the free Euclidean distance of the uncoded reference system (e.g., 4-PSK), and $P_{\text{ave,coded,uncoded}}$ is the average power of the signal for the coded (and uncoded) case.

Clearly, the objective in seeking good codes is to maximize the free Euclidean distance. For high signal-to-noise ratios, the preceding for a given probability of error yields the following expression for coding gain:

$$(E_b / N_o)_{\text{uncoded}} - (E_b / N_o)_{\text{coded}} = \text{Coding Gain} \quad \text{dB} \qquad (6.2)$$

Therefore, in TCM, coding is optimized so as to maximize the minimum free Euclidean distance between possible transmitted signal sequences. This is done by partitioning, as proposed by Ungerboeck.

Rules have been set down by Ungerboeck to establish trellis transition designations. These two rules are generalized from those given by Ungerboeck:

1. When parallel transitions are allowed in the trellis, the two signals to be used are taken from the partitioning shown in Figure 6–3 which gives the *maximum* ED in the partitioned constellation (from Tier (d) in Figure 6–3).

2. All transitions which *diverge* from or *merge* into a trellis state receive the *next* maximum possible ED in the partitioned constellation (corresponds to Tier (c) in Figure 6–3).

These rules will assign signal point transitions in the trellis which will maximize the code free Euclidean distance with a code for each value of constraint length K of the encoder. The value of K determines the number of trellis states.

6.3 Attributes of Trellis Coded Modulation

At this juncture, it is appropriate to present an example of how TCM performance is achieved. Consider the uncoded 4-PSK constellation shown in Figure 6–3(a). This is the reference set for the subsequent *coded* 8-PSK shown in Figure 6–3(b).

Increasing the size of the modulation signal set from $2^n = 2^2 = 4$ to $2^{2+1} = 2^3$ doubles the signal alphabet (as shown). The code used therefore becomes $n / (n + 1) = 2/3$. This results in the *coded* 8-PSK signal shown in Figure 6–3(b). After Ungerboeck set partitioning, the display in Figure 6–3(c) and 6–3(d) results. Notice that as partitioning expands, EDs become larger between signal points, that is, $d_0 < d_1 < d_2$. Note the coded signal minimum, d, is $d_0 = 0.765$. You would think that we have degraded performance because of the reduced distance, but actually, as we have indicated previously, this is not the case. *The coding gain results in the free Euclidean distance, which will be greater than the minimum ED of the reference (4-PSK)*. For the uncoded signal set, d_{free} is simply the minimum ED between signal points.

One possible 8-state trellis (with a constraint length $K = 3$) used by Ungerboeck to manifest a code which maximizes free distance of a signal design is shown in Figure 6–3(e). This trellis does not use parallel transitions, that is, from Tier B0 and B1, and thus adheres to Rule 2 given previously. If n bits (in this case, 2) are to be encoded per modulation interval, there must be $2^n = 2^2 = 4$ possible transitions from each trellis state to a successor state (as shown). Therefore, it is noted that the current state for the example has four transitions and the next state has four entering it (not shown). The numbers at the left of the states in the trellis are symbol design points used for each of the four paths diverging from the state taken from B0 and B1 in Figure 6–3(c).

The uncoded 4-PSK also has a trellis associated with it, but is merely a one-state trellis given as Figure 6–4.

Every connected path represents an allowable signal sequence. Four possible transitions between consecutive nodes correspond to the four possible symbols 00, 01, 10, 11. Note that the free Euclidean distance, $d_{\text{free}} = \sqrt{2}$, is also equal to the ED in signal space.

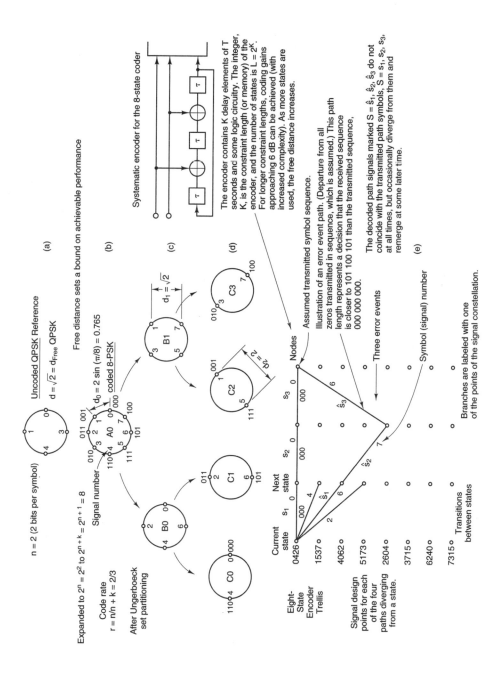

Figure 6–3 Partitioning of 8-PSK signal design into subsets with increasing free subset distance ($d_0 < d_1 < d_2$). Copyright © 1989 by the Institute of Electrical and Electronics Engineers, Inc. Reprinted with permission.

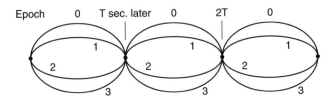

Figure 6–4 Chain trellis.

A demonstration of an error event in the trellis occurs when the decoded symbols do not coincide with the transmitted path symbols, but diverge and reemerge later. This is manifested in the trellis by referring to the assumed *transmitted* symbol sequence, $S = s_1, s_2,$ and s_3. To find the minimum distance of trellis codes, often the all-zeros path serves as the reference path. The departure sequence is $\hat{S} = \hat{s}_1, \hat{s}_2,$ and $\hat{s}_3$. Each separation of S and $\hat{S}$ is an error event. The error events shown in Figure 6–3(e) are designated by the path 6 — 7 — 6. Note that these numbers correspond to the signal numbers on the signal space constellation in Figure 6–3(b), that is, $6 = 101$ and $7 = 100$. The transmitted signal sequence (assumed) is designated as $0 — 0 — 0$.

Therefore, the free Euclidean distance, d_{free}, is the minimum ED between a pair of transmitted code sequences (from the partitioning diagrams in Figure 6–3(b), between points 0 and 6 and 0 and 7), which split at some level in the trellis (top left node) and merge later (at the fourth node). This is the square root of the sum of the squares of the geometric distance between *corresponding* symbols of the two sequences. We therefore have

$$d = (s_1 - \hat{s}_1)^2 + (s_2 - \hat{s}_2)^2 + (s_3 - \hat{s}_3)^2 \qquad (6.3)$$

Note that $s_1 = s_2 = s_3 = 000$, $\hat{s}_1 = 101$, $\hat{s}_2 = 100$, and $\hat{s}_3 = 101$. Therefore,

$$\begin{aligned} d_{\text{free}} &= (s_1 - \hat{s}_1)^2 + (s_2 - \hat{s}_2)^2 + (s_3 - \hat{s}_3)^2 \\ &= (000 - 101) + (000 - 100) + (000 - 101) \end{aligned}$$

From Figure 6–2,

$$\begin{aligned} &= d_1^2 + d_0^2 + d_1^2 \\ &= (0 - \sqrt{2})^2 + (0 - 0.765)^2 + (0 - \sqrt{2})^2 \\ &= 2 + 0.585 + 2 \\ &= 4.585 \\ d_{\text{free}} &= 2.14 \end{aligned}$$

From Equation (6.1), the asymptotic coding gain becomes (assuming average power of the coded and uncoded signals are the same):

$$G_{ACG} = 20\log_{10}(d_{free}/\sqrt{P_{ave}})/(d_{ref}/\sqrt{P_{ave}}) \qquad (6.4)$$
$$= 20\log_{10}(2.14/\sqrt{2}) = 20\log(1.5)$$
$$= 3.52 \text{ dB}$$

Note that d_{ref} is from Figure 6–3(a). The coding gain is therefore 3.5 dB above that obtained from the reference QPSK system.

In the partitioning scheme proposed by Ungerboeck, trellis transitions are assigned to waveforms so that the free Euclidean distance will always exceed the minimum distance of the uncoded reference waveform. In the example above, the reference is the uncoded QPSK.

Increasing the number of states in the trellis improves performance significantly. This is easily done at the transmitter encoder, but the complexity in the system is manifested in the decoder. The complexity of the Viterbi algorithm is roughly proportional to the number of encoder states. However, with the advent of VLSI, this problem is somewhat tempered.

Curves showing the performance of coded 8-PSK for several trellis states are shown in Figure 6–5. The performance for the uncoded QPSK is shown at the right with a $p(e) = 10^{-5}$ for $E_b/N_o = 9.6$ dB.

Nominal gains for various trellis states, which can be achieved using the Ungerboeck scheme, are summarized in Table 6–1. The gains in dB over uncoded signaling for TCM using PSK and 16-ary QAM are shown. Notice that for states above about 64, there is not an appreciable increase in gain. Any higher states may not be worth the complexity and cost. This may be why QUALCOMM has designed a 64-state CODEC device (see Figure 6–9).

Even though we have demonstrated the Ungerboeck concepts for M-ary PSK signals, these schemes may also be applied to M-ary QAM signals and pulse amplitude modulation (PAM). In fact, systems are now available with TCM-coded 256 QAM signals. An example of a partitioned, coded 16-QAM constellation where 8-PSK (3 bits/symbol) is used as the reference is shown in Figure 6–6 [1]. As indicated previously, the EDs increase as the partitioning fans out, in particular, where $d_0 < d_1 < d_2 < d_3$. An 8-state trellis for this constellation is also shown in Figure 6–5. Notice that it has parallel transitions and thus Rule 1, indicated previously, applies. The state designations result from the maximum ED in the partitioned constellation. The free Euclidean distance is $d_f = 1.414$, which has an ACG of 5.3 dB over uncoded 8-PSK.

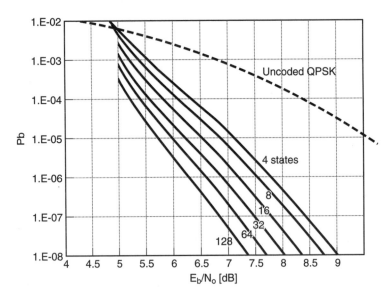

- Coded 8-PSK (2 information bits/T seconds).
- Calculated results.

Figure 6–5 Bit error rate probability vs. E_b / N_o for 8-PSK for trellis codes, rate 2/3 and several states. Copyright © 1989 by the Institute of Electrical and Electronics Engineers, Inc. Reprinted with permission.

Table 6–1 Achievable Gains Using the Ungerboeck Coding Schemes

States	8-ary PSK, rate 2/3	16-ary PSK, rate 3/4	16-ary QAM[*], rate 3/4
4	3.0 dB	3.5 dB	4.4 dB
8	3.6	4.0	5.3
16	4.1	4.4	6.1
32	4.6	5.1	6.3
64	5.0	5.3	6.8
128	5.1	5.4	7.4
256	5.4	5.5	7.4
512	5.7	—	—
1024	5.7	—	—

[*] Relative to uncoded 8-ary PSK. Coding gains can vary depending on the uncoded reference used. A poor uncoded reference constellation can offer a higher gain when compared with a good coded waveform.

Practical Systems 117

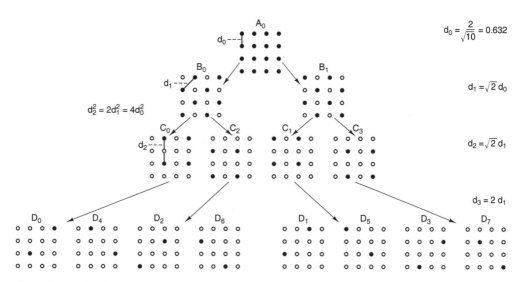

Note each step of partitioning increases the minimum distance by 3 dB.

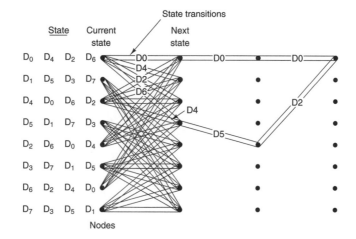

Figure 6–6 Partitioning coded 16-QAM (16-QASK) channel symbols into subsets with increasing minimum subset distances, and associated trellis to find the free distance (and thus asymptotic coding gain). Copyright © 1989 by the Institute of Electrical and Electronics Engineers, Inc. Reprinted with permission.

6.4 Practical Systems

Table 6–2 shows the evolution of telephone line modems. In 1971, modems using higher modulation architectures showed up in the Codex Corporation 9600C. These modems had a 4 b/s-Hz signal efficiency using 16-QAM. They were followed in 1980

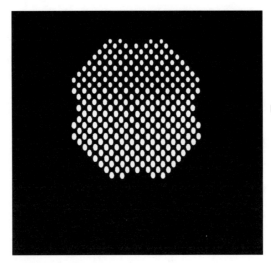

$M = 2^n = 2^8 = 256$

- Uncoded 128-QAM; 7 bits/symbol.
- Encoder adds one bit.
- Uses 8-state trellis.
- Convolutional encoder with constraint length K = 3.

> Trellis coded modulation (TCM) is an extension of QAM which uses error correcting to improve performance. The Codex 2660 TCM modem achieves its error correcting ability by adding redundancy to the transmitted sequence of signal points. Redundancy is introduced by using more signal points in the signal constellation than are needed for the corresponding QAM modem. For example, at 16.8 kb/s, the TCM modem collects the data bits into a 7-bit symbol 2400 X per second and encodes two of the 7 bits through an encoder which adds an extra code bit. The resulting 8 bits are mapped into a signal point selected from a 256-point signal constellation according to the rules of the Trellis Coding Scheme.
>
> *Motorola Inc.*

Figure 6–7 Coded 256-QAM TCM cross constellation. Used with permission of Motorola Inc.

by the Paradyne and Codex SP14.4 systems, which had 6 b/s-Hz efficiency using 64-QAM. In 1984, the Codex 2660 was the first modem to use TCM with 7-bit efficiency with an uncoded signal of 128-QAM. The code added one bit, yielding a constellation of 256 points. A dramatic display of its constellation is depicted in Figure 6–7. An additional increase in speed was achieved in 1985 with the Codex 2680. This was a new concept using an eight-dimensional trellis code with 160-QAM. The use of multi-dimensional codes in the most recent V.34 modems, allow operation of 28.8 kbps and 33.6 kbps.

Practical Systems

Table 6–2 Telephone Modem Milestones

Year	Model	Maximum Bit Rate (bits/s)	Signaling Rate (symbols/s)	Modulation	Signaling Efficiency (bits/symbol)
1962	Bell 201	2400	1200	4-PSK	2
1967	Milgo 4400/48	4800	1600	8-PSK	3
1971	Codex 9600C	9600	2400	16-QAM	4
1980	Paradyne MP14400	14400	2400	64-QAM	6
1981	Codex SP14.4	14400	2400	64-QAM Trellis Coded	6
1984*	Codex 2660	16800	2400	256-QAM	7
1985	Codex 2680	19200	2743	160-QAM 8-D Trellis Coded	7
1993	Codex 3600 V.34	24000			
1994		28800		Multidimensional	
1998	V.90	56000			

* Uses 8-state trellis, $K = 3$ convolutional code.
TCM has made possible the d evelopment of very high-speed modems in voiceband data transmissions.

A TCM coder used in telephone networks is shown in Figure 6–8(a). A redundancy bit is indicated as YO in the block diagram. The signal space diagram for this network is shown in Figure 6–8(b).

Trellis codecs are now becoming commercially available. Figure 6–9 shows a structure, which is available from QUALCOMM, along with its performance. It is interesting to note that this is a complex 64-state trellis, which would be rather difficult to build were it not for VLSI implementation. Both code rates of 2/3 and 3/4 provide a single redundant bit for coding.

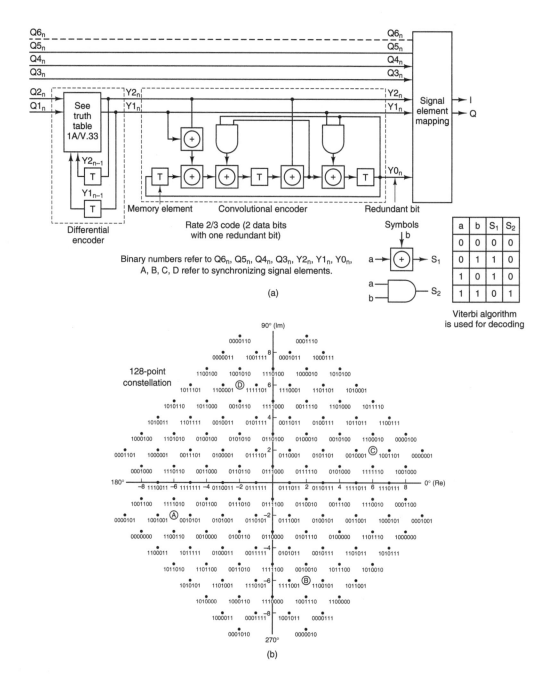

Figure 6–8 (a) Trellis coder at 14,400 bits/sec. (b) Signal state space diagram and mapping for TCM modulation at 14,400 b/s.

Performance Degraders

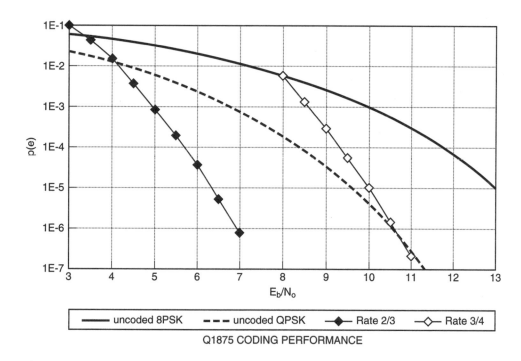

Q1875 CODING PERFORMANCE

Figure 6–9 Commercially available TCM codec, showing performance data.

6.5 Performance Degraders

Several sources of degradation reduce the performance of TCM systems. Of course, these are in addition to Additive White Gaussian Noise (AWGN), which is ubiquitous. These sources include the following:

- Some channels are inherently non-linear (especially satellite channels because of HPAs and TWTAs). If QAM modulation types are used (multi-level signals), system operation must be backed off in power to produce an approximately linear channel. Predistortion is also used to cope with this problem. The use of Solid-state Power Amplifiers (SSPAs) offers better linearity with less backoff.
- Doppler shifts can occur if the platforms on which the modems are located are in motion (e.g., in mobile satellite systems). This must be compensated for to maintain coherent operation.

- Multipath fading or shadowing will make performance erratic if not mitigated. Recall that multipath propagation will assume Rayleigh statistics if a dominant component is not in the signal. With a dominant component, Ricean statistics prevail. Shadowing is a long-time fade effect and usually assumes log-normal statistics.

Some reported results seem to indicate that TCM under deep fade conditions is degraded more than for uncoded waveforms. Therefore, it is more sensitive to perturbations. Clearly, coded M-ary PSK would be less sensitive than coded M-ary QAM. The "jury is still out" in some of these areas. However, the above-mentioned degradations may be substantially overcome by using interleaving over a number of fades, if the inherent interleaving delay is tolerable.

6.6 Conclusions

The following is a summary of the advantages, disadvantages, and uses of TCM:

Advantages:

- Achievable coding gain without incurring the penalty of bandwidth expansion.
- No reduction in data rate since modulation rate does not change because of increase in modulation signal set.
- No increase in transmitted power.
- Realization of coding gains (in minimum distance) up to 6 dB.

Disadvantages:

- More complex demodulator, which increases with the number of trellis states.
- May not offer advantages in multi-fade environment (if interleaving is not used).

Uses:

- High-speed voiceband modems.

- U.S. digital TV standard for HDTV transmission.[2]
- Networks where bandwidth and *power* have constraints.

Until recent years, spread spectrum communications has remained the province of the military and has principally been used for covert radio communications and combatting jamming. As the military relaxed the security requirements, it immediately found applications in the commercial sector in cellular applications. This was exploited as still another method to increase the capacity (additional users) which is confronted by spectral limitations. In Chapter 7 we will consider spread spectrum communications and its attributes.

6.7 Glossary of Terms

ACG: Asymtotic Coding Gain
ASYMPTOTIC CODING GAIN: $G = 20 \log_{10}(d_{\text{free}} / P_{\text{ave}}) / (d_{\text{ref}} / P_{\text{ave}})$ for high signal-to-noise ratio the above becomes
$$G = (E_b / N_o)_{\text{uncoded}} - (E_b / N_o)_{\text{coded}} \quad \text{dB}$$
AWGN: Additive White Gaussian Noise
ASK: Amplitude Shift Keying (one dimensional equally spaced points constellation)
BER: Bit Error Rate
BPSK: Binary Shift Keying
CODE RATE: The ratio per unit time of the *message* bits to total number of bits transmitted.
CODING GAIN: The reduction in E_b / N_o required for providing a specified bit error rate (BER).

---

2. The Advanced Digital Television Standard (ADSC) offers two modes: a *terrestrial broadcast mode* using digital vestigial sideband modulation 8-VSB (with discrete amplitude levels), and a *high data rate mode* using 16-VSB with 16 discrete amplitude levels. Note the signal space constellations for both these modes manifests one-dimensional signal space constellations. The RF/transmissions subsystem also contains FEC (R-S, interleaving, TCM) to combat both bursty errors and random errors. Note that interleaving complements the R-S encoder. TCM is not used in the high data rate mode.

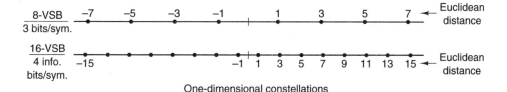

One-dimensional constellations

d MINIMUM: Defined as the minimum Euclidean distance between signal points in signal space constellation, for *uncoded* signal. Sometimes referred to as the reference distance, d_{ref}.

DVSB: Digital Vestigial Sideband (note this is comparable to PAM where there are M-ary discrete amplitude levels.)[3]

ED: Euclidean Distance

ERROR EVENT: Identified as a departure from the all-zeros path followed by a remerging with the all zeros path.

EUCLIDEAN DISTANCE: Distance between signal points in the signal space constellation.

4-PSK: Same as QPSK

FREE EUCLIDEAN DISTANCE, d_{free}: Defined as the minimum Euclidean distance between any pair of transmitted code sequences (which split at some level in the trellis and merge later. The free Euclidean distance between any two code sequences is defined as the square root of the sums of the squares of the geometric distances between corresponding symbols of the two sequences.

GRAY CODE: Single bit difference from one signal symbol to another. For example, 000, 010, 011, 111.

HAMMING DISTANCE: The number of places any two codes differ. For example, 00010 and 11001, $D = 4$. The greater the Hamming distance of a code, the greater is its ability to correct a corrupted message. A larger Hamming distance is achieved by using a more complex code. Note in TCM we deal with Euclidean distances as opposed to Hamming distances.

HAMMING WEIGHTS: Defined to be the number of non-zero components of the code. For example, 011001, $W = 3$.

K : encoder constraint length

M-PAM: M-ary Pulse Amplitude Modulation (one dimensional)

M-PSK: M-ary PSK

M-QAM: M-ary QAM

PAM: Pulse Amplitude Modulation

PARALLEL TRANSITIONS (in a trellis): Results from the transmission of uncoded bits along with coded bits in the convolutional encoder.

PSK: Phase Shift Keying

QAM: Quadrature Amplitude Modulation (two dimensional)

QASK: Quadrature Amplitude Shift Keying (same as QAM)

---

3. The proposed HDTV standard will use 8-VSB and 16-VSB modulations.

QPSK: Quaternary PSK

r: code rate

SOFT DECISION (as opposed to hard decision (yes/no)): A soft decision demodulator first decides whether the output voltage is above or below threshold and then computes a confidence number how far from threshold it is. Soft decision in the decoder, in lieu of hard decision, realizes about a 2 dB improvement in coding gain. Used with the Viterbi decoder in TCM.

TCM : Trellis Coded Modulation (implies that convolutional is used as opposed to block coding, one can implement TCM-like with block coding).

6.8 References

[1] G. Ungerboeck, "Channel Coding With Multilevel/Phase Signals," *IEEE Trans. Information Theory*, January 1982.

[2] ——, "Trellis-Coded Modulation With Redundant Signal Sets," *IEEE Comm. Magazine*, Part I, February 1987.

[3] ——, "Trellis-Coded Modulation With Redundant Signal Sets," *IEEE Comm. Magazine*, Part II, February 1987.

[4] H. K. Thaper, "Real-Time Application of Trellis Coding to High-Speed Voiceband Data Transmission," *IEEE J. of Sel. Areas in Communications*, September 1984.

[5] B. Sklar, "Trellis-Coded Modulation," *Milcom*, October 1989.

[6] G. D. Forney, et al., "Efficient Modulation for Band-Limited Channels," *IEEE J. Sel. Areas in Communications*, September 1984.

[7] G. C. Clark Jr. and J. Bibb Cain, *Error-Correction Coding for Digital Communications*, Plenum Press, New York, 1981.

[8] QUALCOMM Inc., "Single Chip K = 7 Viterbi Decoding Device," *IEEE Communications Magazine*, April 1987.

[9] A. J. Viterbi, "Convolutional Codes and Their Performance in Communication Systems," *IEEE Trans. Communications*, October 1971.

[10] K. Krechmer, "The State of the Modem," *Data Communications*, June 1991.

[11] CCITT Blue Book, "Data Communications Over a Telephone Network," Vol. VIII-Fascicle VIII.1, Melbourne 14-25, November 1988.

[12] IEEE Publication, "Special Issue Coded Modulation," *IEEE Communications Magazine*, December 1991.

[13] IEEE Publication, "Special Section on Combined Modulation and Encoding," *IEEE Trans. Communications*, March 1991.

[14] S. Benedetto, et al., "Performance Evaluation of Trellis Coded Modulation," *Proc. IEEE*, June 1994.

[15] QUALCOMM Inc., "Q1875 Pragmatic Trellis Decoder," *Technical Data Sheet*, May 1992.

[16] E. Zehavi and J. K. Wolf, "On the Performance Evaluation of Trellis Codes," *IEEE Trans. on Information Theory*, March 1987.

[17] B. Pattan, "Spectrally Efficient Higher-Order Modulation Architectures for Terrestrial and Satellite Applications," *FCC/OET Internal Memorandum*, June 10, 1991.

[18] ———, "In Pursuit of Bandwidth Efficiencies for Wireless Terrestrial and Satellite Communications," *FCC/OET Internal Memorandum*, June 1995.

[19] ———, "CODING—Antithesis to Spectral Efficiency," *FCC/OET Internal Memorandum*, August 1995.

CHAPTER 7

Spread Spectrum Communication Systems

7.1 Introduction

Spread spectrum communications, until recently, have resided in the domain of military systems. The advantages of this application are: 1) it has a low probability of intercept because power is spread over a wide bandwidth[1] using a pseudo-random sequence, thus reducing RF-power flux densities, 2) it offers the same measure of security if proper codes are used, and 3) it places a burden on interfering jammers by forcing them to dilute their power over wide bandwidths. Spread spectrum concepts have also found application in radar systems, but in that vein, they are referred to as pulse compression, that is, to place high energy on target to realize required detection probabilities, a long coded pulse is transmitted. Upon reception, the coded pulses are compressed in time to realize range resolution—which a long pulse does not give.

Some of the properties of spread spectrum alluded to above are also applicable to commercial systems. Fortunately, in this area, the atmosphere is usually benign and any interference is known apriori and pallitives are usually employed. One attribute of commercial systems is that many networks using spread spectrum signals may be operating simultaneously. In military situations, the numbers are usually relatively small; a system in fact may be a solo operation. Since there will normally be several systems operating concurrently, in commercial applications there is concern for mutual interference.

---

1. This concept is quite antithesis to present thinking of increasing the spectral efficiency of signals to pack more channels into a limited bandwidth.

Also the fact that the power density emitted by spread spectrum systems is low, the interference to other services will benefit by the low power densities and may not even be "seen" by these other services. On the debit side, spread spectrum systems transmit signals whose bandwidths far exceed the bandwidths of the intelligence information they are transmitting; thus, transmitter-receiver synchronization is required to recover the information signal.

The interest in spread spectrum for use in the commercial sector has increased in recent years and has created much activity in developing systems, including terrestrial applications such as cellular telephone, VSATS, and transmissions via communications satellites. This brief introduction intends to familiarize you with this subject and/or offer greater insight for those already familiar with its theory and technology.

Traditionally, signals have been separated in frequency or time to prevent mutual interference. In spread spectrum systems, the various signals, ironically, use the same frequency, but mutual interference is prevented by the use of unique (orthogonal) codes assigned to each user. Thus, the signals do not "see" each other or mutually interfere. This, of course, is in theory and nothing is perfect. In the real world, there will be some interference to a bona fide user from other users sharing the same spectrum. This is because the codes used are not perfect,[2] and therefore, some interference will seep through. This problem is addressed in so-called Code Division Multiple Access (CDMA) systems, which in military or defense systems is known as the Spread Spectrum Multiple Access (SSMA) technique.[3] CDMA implies multiple access among users in one service, for example, in cellular telephony satellite FFS or VSAT use. There are also other services (terrestrial) which may be transmitting in the same band, and of course, these will attempt to negotiate the receiver. If the bogus signal is not matched to the receiver (spread spectrum), discrimination will be offered against the intruder.

Spread spectrum implies a modulation technique in which the message information, which resides in a narrowband spectrum, is modulated onto a vehicle signal which has a larger bandwidth, resulting in a composite signal whose RF bandwidth is nonetheless equal to the RF bandwidth of the vehicle signal. Since the signal has been spread over a large bandwidth, the power density at any one frequency in this wideband is reduced appreciably. This is why spread spectrum signals have low detectability—

---

2. The unique codes are not quite orthogonal to each other. More will be said on this in later sections.
3. Other modes of access include Frequency Division Multiple Access (FDMA), of which SCPC is a form, and Time Division Multiple Access (TDMA). The former divides the spectrum among the users; the latter, a user is assigned exclusive use of a time slot. Notice that signals, in both of these access techniques, are orthogonal. Other multiple access schemes include Polarization Division Multiple Access (PDMA) and Space Division Multiple Access (SDMA), which is a frequency reuse scheme.

they can reside below the noise floor. Due to this fact, in military circles, this is usually referred to as a low probability of intercept, or LPI. In fact, the signal is buried in the noise. It takes a special receiver to extract it, operation of which is possible due to large realizable processing gain.[4]

7.2 Spread Spectrum Techniques

There are basically three general forms or types of spread spectrum techniques. They are:

1. Direct sequence.
2. Frequency hopping.
3. Hybrid of the two.

Chirp signals, or FMing, have also been used to increase the bandwidth of a signal, but these types have mostly been used in radar systems using pulse compression

We will consider these various spreading techniques in subsequent paragraphs. We will also discuss coding techniques instrumental in spreading spectrum, interference situations and benefits accrued by spread spectrum, and the operation of several spread spectrum systems in assessing the transponder of a satellite.

Widening the transmitted spectrum to improve the S/N ratio is not a new concept. FM signaling is, in a sense, a spread spectrum signal since a wideband signal is transmitted. Recall that the output S/N ratio and FM processing gain (G) are given by

$$(S/N)_{out} = 3/\beta^2 (S/N)_{in} = G(S/N)_{in} \qquad (7.1)$$

where $(S/N)_{out}$ is the output S/N ratio, β is modulation index, $(S/N)_{in}$ is the input S/N ratio, and G is the FM processing gain.

However, an FM signal differs from a spread spectrum signal in a very fundamental way. In spread spectrum, the information signal is not itself processed to enhance the bandwidth, but an independent signal acts as this vehicle. Processing gain is obtained from RF spreading during the spread spectrum process. In FM, gain is realized by utilizing the intelligence of the signal itself, which is inherently spread.

The basic components used in spread spectrum systems are depicted in Figure 7–1. The top figure shows a digital technique where the spreading function is a pseudo-random digital pulse train. This will be explained further in a later section. The second configuration is a so-called frequency hopper, in which the output is a set of one-at-a-time

---

4. The processing gain can be compared to the FM advantage, a feature associated with FM systems.

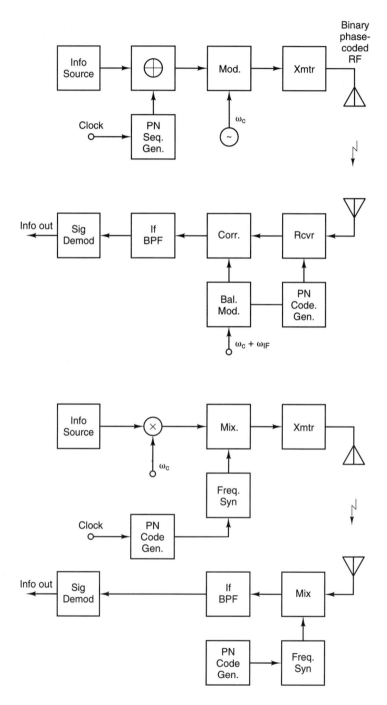

Figure 7–1 Direct sequence and frequency hopping spread spectrum functional block diagrams.

discrete frequencies randomly spread over a wideband. Both use a pseudo-random code generator to produce the side spectrum. The digital system uses the PRC bitstream mod-2 added to the information. The PRC in the frequency hopper generates random frequencies. In the direct sequence approach, the output is a sequence of 0 and 1 pulses which are pseudo-random, and its frequency domain representation is a very wide spectrum.

7.2.1 Direct Sequence Concept

In the direct sequence concept, both the information signal and a spreading signal are in digitized format. Since these are in digital form, they may be combined using modulo-2 addition. Modulo-2 addition is also referred to as an exclusive-or circuit. In this technique, if two binary digits being added are the same (ones or zeros), the added output is "zero", if the two bits are different (a one and a zero), the output is a "one". This operation is indicated in Table 7–1 below.

Table 7–1 Modulo-2 Addition of Information Bit and Coding Bit

Information Bit		Spreading Code Bit	Resultant Sequence
1	$\oplus^*$	0	1
1	$\oplus$	1	0
0	$\oplus$	1	1
0	$\oplus$	0	0

* $\oplus$ represents modulo-2 addition; also referred to as an exclusive-or gate.

The information bit rate is normally much lower than the spreading signal bit rate, and the ratio of the two determines the extent to which the spectrum is spread. The larger the ratio, the wider the spectrum is spread and the larger the processing gain. A pseudo-random code is used to generate the spreading function. Pseudo-random code is used so that the transmitted spread spectrum signal will appear as noise-like as possible, that is, a nearly uniform power spectrum with correlation functions that are impulsive. Actually, the power density spectrum and auto-correlation function are related via the Weiner-Khintchine Theorem, that is, the power spectral density and auto-correlation function are Fourier cosine transform pairs.

Thus,

$$R(\tau) = (1/2\pi)\int_{-\infty}^{\infty} S(\omega)\exp(j\omega\tau)d\omega \qquad (7.2)$$

$$S(\omega) = \int_{-\infty}^{\infty} R(\pi)\exp(-j\omega\tau)d\tau \qquad (7.3)$$

Because the auto-correlation function is a real function, the relationships can be expressed as

$$R(\tau) = (1/\pi)\int_{0}^{\infty} S(\omega)\cos\omega\tau d\omega \qquad (7.4)$$

$$S(\omega) = 2\int_{0}^{\infty} R(\tau)\cos\omega\tau d\tau \qquad (7.5)$$

If the spread spectrum were purely white (uniform power spectrum), it would appear in the frequency domain as

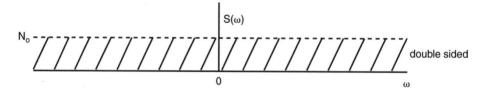

Therefore, for a white noise spectrum

$$R(\tau) = (1/2\pi)\int_{-\infty}^{\infty} N_0 \exp(j\omega\tau)d\omega = N_0\delta(\tau) \qquad (7.6)$$

A white spectrum (like white noise) has a correlation function which is a delta function, or spike at $\tau = 0$.

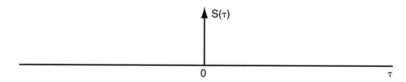

The variance of a white spectrum, σ^2, is infinite because $R(0) = \sigma^2$, and for a white spectrum, the impulse at the origin has infinite height.

In actuality, the spectrum of a spread signal is not really uniform, but is finite and has a $\sin^2 x/x^2$ spectrum distribution; the correlation function is not impulsive, but has finite width and appears triangular as indicated below.

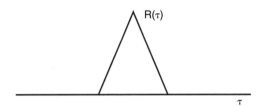

The spectrum is therefore noise-like, but not perfectly white. Clearly, an impulsive $R(\tau)$ suggests infinitive width of the spectrum—a mathematical fiction.

At the receiver, a replica of the coded waveform is used to generate the received spread spectrum signal to remove the coded waveform from the ss signal and produce the transmitted intelligence signal. The de-spreader is basically a correlator which correlates the input signal with a local reference signal. The receiver will pass a signal which is matched to its locally generated code and reject (not entirely) signals which do not have matching codes. These extraneous signals may include many spread spectrum signals, and even relatively narrowband classical signals. Ideally, the receiver would reject all signals but those from its parent transmitter. The extraneous incident spread spectrum signals will remain smeared and appear as low spectral density signals to the receiver. The narrowband signals will be spread, thus diluting their impact.

7.2.2 Frequency Hopping Concept

In a frequency hopping system, the signal transmitted is a series of spectral spikes that are received at random locations in the RF band. At the receiver, synchronization (signal overlap) is such that the local code will generate a spectral spike which matches the incident spike and thus remove that frequency. This recovers the information signal. Notice that in both concepts, a pseudo-random generator produces the signal that is combined with the intelligence signal to be transmitted.

7.3 Code Generation

Generation of codes which have good random properties is an important requirement in spread spectrum systems. The codes are used to spread the spectrum, and hopefully, the spectrum generated will appear as noise-like as possible. The reception of this randomized signal, with the embedded intelligence, is processed by a de-spreader, or correlator. De-spreading is ideal if the spread signal manifests a correlation function which is single valued with small correlation peaks when the received signal and correlator reference are not coincident, that is, at coincidence, the two spreading codes "line up" and the spread signal is removed from the intelligence signal.

A common method of generating pseudo-random sequence, which is used to spread the spectrum, is by using shift register generators. The codes are called pseudo-random because the stream of bits in the bit output is known apriori and the code is established by a deterministic mechanism, that is, linear shift registers of finite length. In a *purely* random sequence, the next bit in the stream is not known apriori. Purely random sequences are not useable in spread spectrum since both transmitter and receiver must know what each succeeding bit will be, and they are of an infinite length, thus making synchronization impossible.

An example of a simple shift register using three stages is shown in Figure 7–2. More practical shift registers use many more stages. The more stages used, the more noise-like the spectrum appears since lines in spectrum are closer. The output of the shift register is initiated by clock pulses, and each clock pulse causes one bit to be dumped to the output. The separation between the clock pulses establishes the width of the pulse (0 or 1) output. The greater the repetition rate of the clock pulses, the shorter the output pulses. These output pulses are commonly called chips.

For the register shown in Figure 7–2, the length of the output sequence is $2^n - 1$, where n is the number of stages. The length is therefore 7 and thereafter the sequence repeats. The initial loading (preset) of each register (basically flip-flops) is "one", and each clock pulse moves the state of the register down the chain.

The loading of each stage with the advent of each clock pulse is indicated in Figure 7–2(a). A 7-bit sequence is generated by this shift register and then repeats itself, 11101001110100.... These sequences are called maximal-length sequences. The sequences are modulo-2 added to the information bits that produce a denser bitstream, which reflects the spread spectrum when viewed in the frequency domain. This operation is shown in Figure 7–3.

The maximal-length sequences alluded to above possess ideal correlation functions. The auto-correlation function of a binary sequence, defined as $R(\tau)$, equals the number of bit agreements minus the number of bit disagreements when the sequence is compared with a cyclic shift of itself. For a zero shift, there is total agreement and thus the correlation value is equal to the length of the bitstream, or $2^n - 1$. For any bit delay increments ($\tau \neq 0$) between sequences, the correlation value will be equal to -1. The correlation functions of maximal-length sequences are therefore two-valued. Symbolically,

$R(\tau) = 7$ for $\tau = 0$ (sequences lined up = period of sequence)
$R(\tau) = -1$ for $\tau = 1, 2, ..., 6$

Code Generation

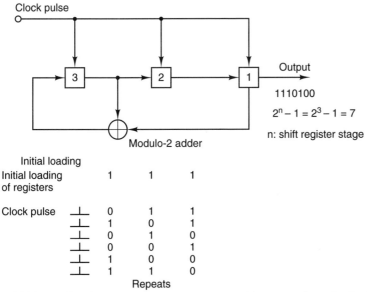

(a) Maximal-length, or m-sequences, are generated by n-stage linear feedback shift registers. These sequences have period "p" equal to $2^n - 1$, where "n" is the number of stages. Practical structures have many stages.

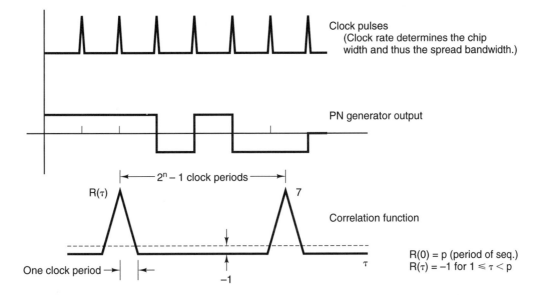

(b) Periodic autocorrelation function of m-sequence.

Figure 7–2 PN sequence generation via a simple shift register.

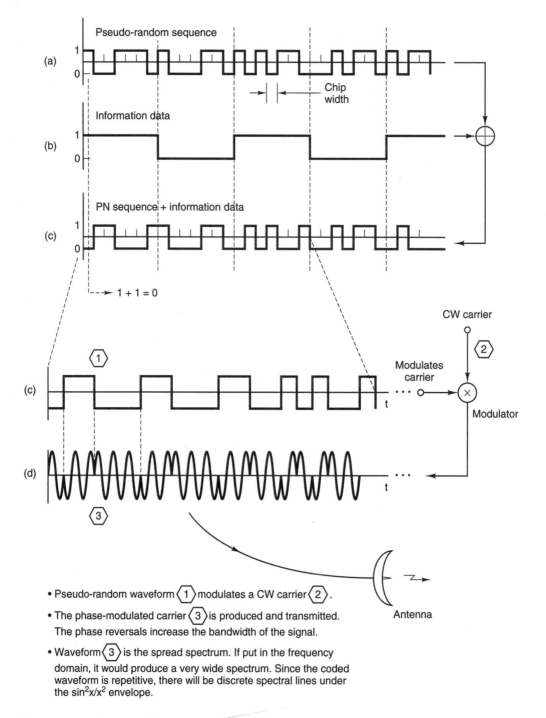

Figure 7–3 Mixing of PRN sequence and data stream at baseband, and generation of RF carrier (coded) for transmission.

Code Generation

The correlation function is shown in Figure 7–2(b).

The triangular waveforms require some explanation. When the sequences are in time coincidence, the apex of the trianpher represents perfect correlation. As the second replica sequence slides over in time (it may shift left or right), the correlation falls off linearly, indicating less and less correlation until the two chips cease to overlap and the correlation is minimal. Mathematically this may be represented as follows:

$$R(\tau) = \begin{cases} 1 - |\tau|\dfrac{2^n}{2^n - 1} & \text{for } |\tau| \leq 1 \text{ chip width} \\ -\dfrac{1}{2^n - 1} & \text{for } |\tau| \geq 1 \text{ chip width} \end{cases} \quad (7.7)$$

Note that these relationships are normalized. When $\tau = 0$, $R(\tau) = 1\left(\frac{7}{7}\right)$, and $\tau \geq 1$, chip width $R(\tau) = -1/2^n - 1$.

As mentioned previously, the Fourier transform of the correlation functions yields the power spectral density. The power spectrum of the auto-correlation function of the pseudo-random code shown in Figure 7–3(c) is given by:

$$S(\omega) = \frac{p+1}{p^2}\left[\frac{\sin(\omega\delta/2)}{\left(\frac{\omega\delta}{2}\right)}\right]^2 \sum_{\substack{m=-\infty \\ m \neq 0}}^{\infty} \delta\left(\omega - \frac{2\pi m}{p\delta}\right) + \frac{1}{p^2}\delta(\omega) \quad (7.8)$$

where p is the period of sequence (number of chips) and δ is the chip width.

Clearly, this signal is not transmitted, but must modulate a carrier. It is recalled from modulation theory that a multiplication (modulation) in the time domain is equivalent to a shift of ω_c in the frequency domain. We therefore have (which also includes the hidden digitized information).

$$S'(\omega) = \frac{p+1}{p^2}\left[\frac{\sin\left(\frac{\omega-\omega_c}{2}\right)\delta}{\frac{(\omega-\omega_c)\delta}{2}}\right] \sum \delta\left[(\omega-\omega_c) - \frac{2\pi m}{p\delta}\right] + \frac{1}{p^2}\delta(\omega-\omega_c) \quad (7.9)$$

A sketch of this spectrum is depicted in Figure 7–4. A spectrum analyzer display is shown in Figure 7–4(b).

Figure 7–5(b) shows the spectrum of a direct sequence spread spectrum signal. There are also discrete spectral lines under the lobes, but they are so close that we are

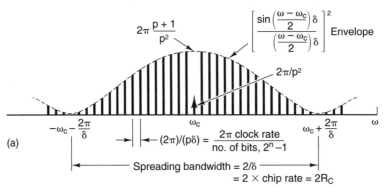

(a)

where p: length of sequence = $2^n - 1$.
  δ: chip width (or bit width) in sequence.
  ω_c: carrier.
  $1/\delta$: clock rate (the faster the rate, the narrower the chips).

Note: • As "p" and/or "δ" $\to \infty$, spectrum granularity becomes continuous.
  • Nulls are separated by twice the chip rate.
  • The number of lines under the main lobe is equal to two times the number of bits (chips) in the spreading code. For example, for the 7-bit code given previously, the number of lines is 14 (does not include same spike at ω_c).
  • The 3dB bandwidth = 0.433 × null-to-null bandwidth.

(b)

Figure 7–4 Power spectrum of maximal-length sequence.

not able to discern them.[5] The bandwidth of this spread signal is measured between the nulls of the main lobe. This width is dictated by the chip width rate of the code used to modulate the information digital signal, which is equal to twice the chip rate. A chip is a simple element of the digital waveform in the spreading code.

Code Generation

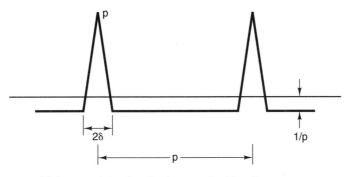

(a) Autocorrelation function for a maximal-length sequence.

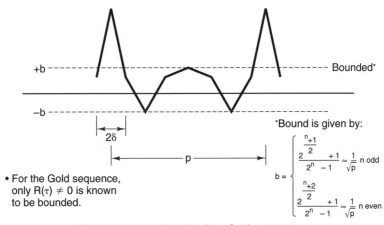

- For the Gold sequence, only $R(\tau) \neq 0$ is known to be bounded.

*Bound is given by:

$$b = \begin{cases} \dfrac{2^{\frac{n+1}{2}}+1}{2^n-1} \approx \dfrac{1}{\sqrt{p}} & n \text{ odd} \\ \dfrac{2^{\frac{n+2}{2}}+1}{2^n-1} \approx \dfrac{1}{\sqrt{p}} & n \text{ even} \end{cases}$$

(b) Autocorrelation function for a Gold sequence.

Note: Correlation codes should possess sharply peaked autocorrelation functions and low cross-correlation peaks. The former facilitates synchronization at the receiver, and the latter prevents locking on a spurious signal.

Figure 7–5 Autocorrelation function of PN sequences.

Because the autocorrelation function is repetitive, the spectrum consists of discrete lines which are separated by $1/p\delta$. The last term in the right member of the equation above is a dc-term and is due to the imbalance in the +1s and –1s in the sequence. It should be noted that if p is doubled, the lines under the sine function are also dou-

---

5. The spectral lines are separated by the period of the pulse sequence used to spread the spectrum. If the period sequence (many chips) is long, the spectrum is essentially a continuous spectrum, or a continuum. As the density of the lines increases ad infinitum, the power in each line is reduced to maintain the conservation of energy.

bled, but reduced in magnitude by a factor of two (there is only so much energy). As we might expect, increasing p still further produces a spectrum which approaches a continuum. It is also noticed that as $p \to \infty$, the dc term gets smaller. If the chip length is halved, the spectrum spreads out to twice its size.

The bandwidth of direct sequence signals is almost always given as the band between the first nulls on either side of the carrier frequency, ω_c. Usually the bandwidth is given as the band between the −3 dB points. The null-to-null bandwidth may then be converted to 3 dB bandwidth by multiplying the null-to-null bandwidth by 0.443, $(2/\delta)0.443$, or $0.886\ Rc$. Clearly this is for the sinx/x type spectrum.

A spread spectrum signal contains 90% of its total power in the main lobe. The balance is in the spectral sidelobes. The power loss in the spread spectrum due to spectral truncation of the signal is indicated in Table 7–2.

Table 7–2 Spectral Power Distribution

Lobe	Bandwidth (2x chip rate)	% of Total Power	Power Loss in dB
main	$2R_c$	90.0	0.45
1st	$4R_c$	94.8	0.23
2nd	$6R_c$	96.4	0.16
3rd	$8R_c$	97.3	0.12
4th	$10R_c$	97.8	0.10

7.4 Codes for Spread Spectrum Multiplexing

Ideally, for CDMA operation of spread spectrum signals, mutual interference can be mitigated by using orthogonal codes to spread the spectrum. Two sequences are orthogonal if the degree of correction approaches zero over the entire sequence. Codes satisfying this requirement are not limited. If they were, there would not be any correlation noise to enter a non-matched receiver and degrade its performance. The number of spread spectrum systems could be unlimited.

Maximal-length sequences are *generally* not suitable for use in CDMA since their *cross-correlation* produces relatively high spurious peaks which the receiver may erroneously interpret in lieu of the true correlation peak. However, there are *select* m-sequences which are usable for this application, but this number is limited.

Other codes which generally have good autocorrelation and cross-correlation functions are the so-called Gold codes. These are also pseudo-random sequences. Gold codes are not orthogonal codes, but display correlations which are *bounded*. This cannot be said of maximal-length codes. Maximal-length codes have better auto-correlation functions than Gold codes since the former are two-valued and have low $R(\tau) = k$, for $\tau \neq 0$, whereas the Gold codes have higher side peaks of their auto-correlation function. An example of these correlation functions is shown in Figure 7–5. The merit of Gold codes is that the sidelobes for both the correlation and cross-correlation functions are bounded, and generally, the maximal-length codes can have high cross-correlation sidelobes.

Being able to find codes in which the cross-correlation peaks are bounded is an important fact to know in CDMA systems. The upper bounds on the peaks of a Gold sequence are given by Gold as

$$\max R(\tau) \leq 2^{(n+1)/2} + 1 \quad \text{for } n \text{ odd}$$
$$\leq 2^{(n+1)/2} + 1 \quad \text{for } n \text{ even} \quad (7.10)$$

Gold codes are generated by modulo-2, combining the outputs of two preferred maximal-length sequence generators.[6] The feedback types are then arranged according to an n-th degree primitive polynomial. The output is a Gold code. The configuration is shown in Figure 7–6. An n-stage MLS generator may assume several different configurations giving different maximal-length sequence outputs. For example, and $n = 5$ stage generator can assume six different forms, depending on how the internal modulo-2 addresses are arranged among the stages. Using narrow combinations of these sequences yields different Gold codes that can be generated from one pain of maximal-length sequence generators, $2^n - 1$. For a 5-stage register, 31 Gold codes can be generated.

Table 7–3 shows that the number of Gold codes generated can be quite large. The table shows the correlation properties of preferred Gold codes. As an example given by Gold, for a shift register containing 13 stages, the maximum number of available MLSs is 630. There are pairs of sequences which will give cross-correlation values as high as $R(\tau) = 703$. But, if selected in accordance with the "Gold Theorem", the correlation value satisfies the inequality indicated previously.

$$|R(\tau)| \leq 2^{(13+1)/2} + 1 = 129 \quad (7.11)$$

---

6. Not the only way to generate Gold codes.

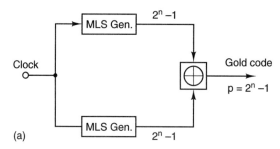

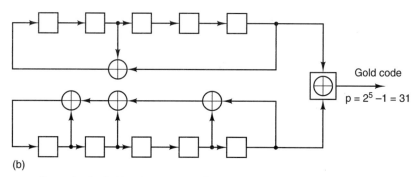

- Two preferred pairs of MLS generators will form $2^n - 1$ Gold codes.
- Generally, Gold codes are not maximal-length sequences, thus their autocorrelation sidelobes will be larger than for MLSs, but their sidelobes are bounded (see Figure 7.5)

Example of a Gold code generator. Thirty-two sequences can be produced by shifting one MLS sequence bitstream one bit at a time.

Figure 7–6 Gold code generation.

The ratio of 129 to maximum correlation value 8191 yields 0.016. These figures are shown in Table 7–4.

Table 7–4 also shows the ratio of correlation to cross-correlation values. Clearly this is important in multiplexed or CDMA systems. It is also obvious from the table that small numbers of register stages provide little immunity of receivers to other spurious codes because of their high correlation peaks which can trigger false responses. The receiver thus could lock on to the wrong code signal.

In Column 3 of Table 7–3, even though the number of sequences is large, they may not all be suitable for the Gold code since they manifest high values of cross-correlation function. A quick test is to determine if the sequences are primes or contain a small number of prime factors. For example, for $n = 5$, the sequence length is 31. This is a prime sequence. Additional prime sequences in the table are 3, 5, 7, and 13. Others

Table 7-3 Properties of Gold Codes*

Number of Shift Register Stages	Spreading Code Length	Number of Available Maximal-Length Sequences	Number of Gold Code Families	Number of Gold Code Sequences in Each Family	Total Number of Gold Codes
3	7	2	1	7	7
4	15	2	1	15	15
5	31	6	15	31	465
6	63	6	15	63	945
7	127	18	153	127	19,431
8	255	16	120	255	30,600
9	511	48	1128	511	576,408
10	1023	60	1770	1023	1,810,710
11	2047	176	15,400	2047	31,523,800
12	4095	144	10,296	4095	42,162,120
13	8191	630	198,135	8191	1,622,923,785

* Many Gold codes can be generated from a relatively few shift register stages. For cellular radio or satellite applications, it would appear that registers with stages in the 4–6 range may be adequate.

not shown give prime sequences for $n = 17, 19, 31, \ldots$. The sequences shown have several prime factors and are still acceptable if they manifest reasonable cross-correlation sidelobes. The fewer prime factors the better, in other words.

When you have two sequences with different lengths, the shorter one is the better because it has lower cross-correlation values, e.g., sequences $2^{11} - 1 = 2047$ and $2^{12} - 1 = 4095$. The first sequence has two prime factors, 23 and 29. The second has four prime factors: 3, 5, 7, and 13. Thus, the first has fewer prime factors and better cross-correlation sidelobes (smaller). Table 7–5 is a list of Mersenne primes. Sequence lengths given by the prime values of n give the lowest cross-correlation values and are best in CDMA operation.

Table 7–4 Correlation Properties of Preferred Gold Codes

Number of Stages in Each Shift Register Generator	Maximum Cross-Correlation Value for Preferred Gold Codes	Correlation Value for Desired Code	Ratio of Cross-Correlation Value Derived by Correlation Value of a Desired Code
3	5	7	0.71
4	9	15	0.60
5	9	31	0.29
6	17	63	0.27
7	17	137	0.13
8	33	255	0.13
9	33	511	0.065
10	65	1023	0.064
11	65	2047	0.032
12	129	4095	0.032
13	129	8191	0.016
14	257	16,383	0.016
15	257	32,757	0.0078
16	513	65,535	0.0078
17	513	113,071	0.0039
18	1025	262,143	0.0039
19	1025	524,287	0.0020
20	2049	1,048,575	0.0020

Table 7–5 Known Mersenne Primes, July 8, 1987. (From *Scientific American*, "The Search for Prime Numbers," by Carl Pomerance.)

Value of n for which $2^n - 1$ is prime	$2^n - 1$
2	3
3	7
5	31
7	127
13	8191
17	131071
19	524287
31	2147483647
61	19 Digits
89	27 Digits
107	33 Digits
127	39 Digits
521	157 Digits
607	183 Digits
1279	386 Digits
2203	664 Digits
2281	687 Digits
3217	969 Digits
4253	1281 Digits
4423	1332 Digits
9689	2917 Digits
9941	2993 Digits
11213	3376 Digits
19937	6002 Digits
21701	6533 Digits
23209	6987 Digits
44497	13395 Digits

7.5 Spread Spectrum Interference Analysis

7.5.1 One-on-One

An interfering signal incident on a spread spectrum receiver will have a spread signal because of the correlation action. The interfering signal will therefore be smeared, at least over the spread spectrum bandwidth. Since this process involves multiplication of the local reference with the incident interfering signal, the interfering will be spread to a greater bandwidth than the spread frequency since there is a convolution of the two spectra ($B = B_{ss} + B_I$). The power spectral density (I_o) of the interferer will be low due to its spreading over the wide bandwidth, and can be assumed to be uniformly distributed.

$$I_o = I + B \quad \text{watts/Hz} \tag{7.12}$$

where I is the total power in the interfering signal and B is the bandwidth of the spread interfering signal.

Frequently in interference analysis it is assumed that the interfering noise (I) is greater than the receiver thermal noise (N). That is,

$$N < I$$
$$\text{or } N_o < I_o \tag{7.13}$$

Figure 7–7(a) shows a model of a single interferer and bona fide signal going through the receiver, in particular, the band-pass filter for the information signal. Even though the interference is spread over a wide bandwidth, only that portion which is within the filter bandwidth will interfere. Note that the smeared interference may be either a narrowband signal which has been smeared or another spread spectrum signal which remains smeared since it is not matched to the receiver.

In this interference environment, the E_b / N_o after the correlation process is

$$E_b / N_o = (P / I)(B / R_b) \tag{7.14}$$

where P is the received bona fide signal power which has entirely negotiated the correlator, R_b is the bandwidth of the information signal in information bits/second, B is the spreading bandwidth, I is the total interference noise (N_o assumed small), and E_b / N_o is required to realize desired BER.

In terms of interference power to receiver signal power

$$I/P = \frac{B/R_b}{E_b/I_o}$$
$$\text{or } I/P = (B/R_b) - (E_b/I_o) \quad \text{dB} \tag{7.15}$$

Spread Spectrum Interference Analysis

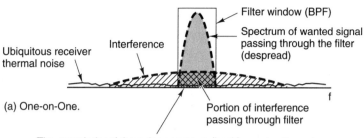

(a) One-on-One.

The granularity of the noise spectrum (hash) passing through the filter is a function of the sequence length. For short sequences, the spectrum will appear spiky and not necessarily noise-like. Short sequences have limited utility.

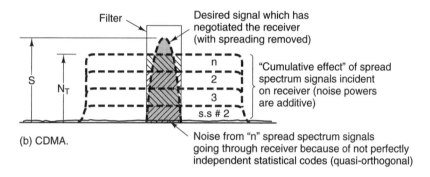

(b) CDMA.

Note: The interference spectra are <u>not</u> continuous, but noise-like. They consist of discrete spectral lines and their density is determined by sequence length and chip width. The longer the sequence, the more interference spectral lines fall inside the band-pass filter. For long sequences, these lines are extremely closely spaced, and for all practical purposes, can be considered a continuum, as shown in Figure 7-4.

Figure 7–7 Interference cross-correlation noise (hash) negotiating the receiver to the output.

The ratio of B / R_b is referred to as the processing gain. This is also equal to the spreading bandwidth to information bandwidth ratio. Notice that the greater this ratio the *greater* the interference to receive a bona fide power ratio that is acceptable and still maintains performance.

We can make the following observations from Equation (7.15):

- Communication for the required BER probability is *not* possible if the following is true:

$$(I / P) > (B / R)/(E_b / N_o) \qquad (7.16)$$

- The desired performance is feasible if

$$(I/P) < (B/R)/(E_b/N_o) \qquad (7.17)$$

Figure 7–8 shows a plot of processing gain (PG) versus the interference to signal power ratio for parametric $p(e)$ probabilities. Clearly, the higher the processing gain, the higher the interference to signal ratio which can be tolerated for the same $p(e)$. This is merely the plot of Equation (7.15) where I/P is referred to as the interference margin.

It should be noticed that if the bona fide E_b/N_o ratio is reduced by coding, there is an additional improvement in interference margin.

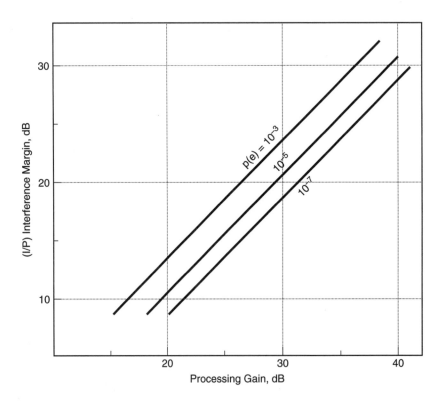

Figure 7–8 Interference to power ratio as a function of processing gain.

7.5.2 Multiply Access Interference Scenario

In a CDMA scenario, there is a multiplicity of spread spectrum networks operating simultaneously and in proximity to each other. The various signals offer potential

Spread Spectrum Interference Analysis

interference to each other. The signals discriminate against each other by employing unique nearly orthogonal codes, and all manifest some degree of orthogonality with respect to each other. It is essential that at least quasi-orthogonality prevails to prevent tolerable mutual interference. The maximum number of simultaneous transmissions is limited by the generation of the cross-correlation noise occurring between wanted and unwanted pseudo-random codes, (because they are not perfectly orthogonal). This interference situation will be discussed here. A scenario depicting multiple access to a receiver is shown in Figure 7–9.

The introductory analysis presented a one-on-one situation where a single interferer plus the wanted signal was accessing the receiver (only one is acceptable). Now assume there are several spread spectrum signals, including the desired signal, of equal power and bandwidth incident on the receiver. Equal power is accomplished by automatic power control. The total power in the ensemble of signals is

$$P_T = \sum_{i=1}^{M} P_i = P_c \tag{7.18}$$

Since one is the desired signal, the signal power, P_b, is therefore

$$P_b = P_c/M \tag{7.18a}$$

where M is the total number of accesses and P_c is the total power. The total power in the interfering signals (not matched to the receiver) is clearly

$$I = \left[\frac{M-1}{M}\right] P_c \tag{7.19}$$

If the bandwidth of the spread spectrum signal is B, the interference noise spectral density is

$$I_o = IB = \left[\frac{M-1}{M}\right] \frac{P_c}{B} \tag{7.20}$$

If thermal noise is added, the composite noise is

$$I_o' = I_o + N_o \quad \text{(where previously we indicated } N_o < I_o\text{)}$$

At the output of the correlator's narrow band-pass filter, which allows the desired despread signal spectrum to pass, the output signal spectral density is

$$P_o = P_b/b = (P_c/M)(1/b) \tag{7.21}$$

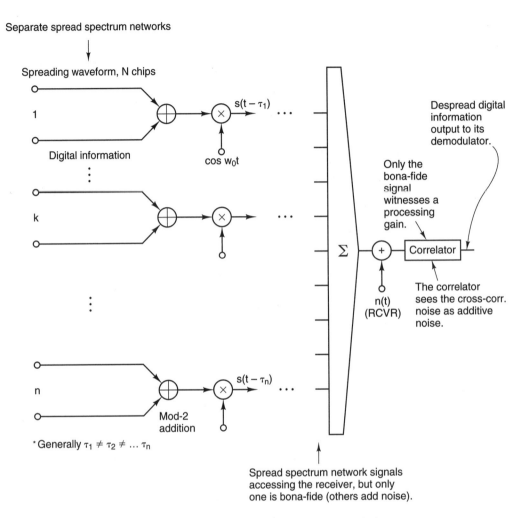

Figure 7–9 Code Division Multiple Access (CDMA) interference environment.

Spread Spectrum Interference Analysis

where b is the post-correlator band-pass filter passing the information signal. Introducing the equation for digital communication, we have

$$(P_b / N_o) = (E_b / N_o)R \tag{7.22}$$

where P_b is P_c / M = true signal power received, N_o is the receiver thermal noise power spectral density, E_b / N_o is the required dB-Hz to give $p(e)$ desired, and R is the capacity of system in bits/sec. We previously assumed that $I_o \gg N_o$, which is frequently the interference environment, and plugging I_o into Equation (7.22) in place of N_o and $P_b = P_c / M$ from Equation (7.18), we have

$$\frac{P_c/M}{\left[\frac{M-1}{M}\right]\frac{P_c}{B}} = (E_b/N_o)R$$

$$\frac{B}{(M-1)} = \left[\frac{E_b}{N_o}\right]R \tag{7.23}$$

Therefore, to communicate with the desired BER

$$M - 1 < (B / R) / (E_b / N_o) \quad ^7 \quad \text{Active users}$$
$$\text{or} \quad (E_b / N_o) = (B / R) / (M - 1) \tag{7.24}$$

If voice activation and customer traffic density is considered (measured in Erlangs), he number of potential users becomes

$$N' = \frac{\text{number of users per cell} \times \text{activity factor}}{\text{traffic intensity (Erlangs)}} \tag{7.24a}$$

We notice from Equation (7.24) that for B / R fixed, $(M - 1)$ is the maximum number of interferers which can be incident on the receiver before the $p(e)$ is increased through the reduction in the required E_b / N_o. For example, $p(e) = 10^{-5}$ for coherent PSCK, a processing gain of 1000, and no coding. In this case, the E_b / N_o required is 9.6 dB-Hz, and the maximum number of users from Equation (7.24) which can exist before deleterious interference noise is significant is $M = 110$ users. For a $p(e) = 10^{-6}$ and using CPSK, the E_b / N_o required is 10.5 dB-Hz. Now the maximum, M, is reduced to 90. Therefore, for performance which requires small $p(e)$s, the number of operating networks is reduced. The number of users can be increased by increasing the processing gain (less correlation noise into the receiver) and/or by reducing the E_b / N_o required to

---

7. This does not include benefits accrued from using voice activation (DSI) and/or customer traffic density (Erlangs).

realize the same $p(e)$. Expressed differently, the effects of interference from multiple users can theoretically be reduced as much as desired by increasing the processing gain. The limit here, of course, is when the cross-correlation noise level approaches that of the receiver thermal noise, where it then sets the absolute limit on performance.

7.6 The Multipath Phenomenon

In the real world, the EM wave to the receiver consists of several paths (multipath propagation). The important path is the direct path, while the non-direct paths, i.e., the sum of all other stray signals, stem from reflections off terrain, buildings, and other objects.

The model for the multipath condition is indicated in Figure 7–10.

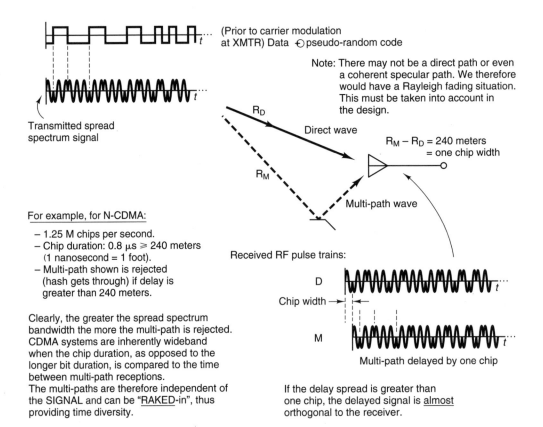

Figure 7–10 Multipath rejection by spread spectrum communications.

The Multipath Phenomenon

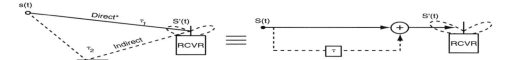

* Direct path is flexing reality in practical mobile terrestrial systems. Generally, the signal loss falls off greater than the range inverse factor, 1 / R2, or 6 dB/octave. Actually, it is more like 1 /R3, or 9 dB/octave. Direct path is more common in satellite networks.

$$s'(t) = s(t - \tau_1) + s(t - \tau_2) \qquad (7.25)$$

where $\tau_1 < \tau_2$, τ is the delay of multipath signals (s).

$$s'(\omega) = s(\omega)\exp(-\omega\tau_1) + s(\omega)\exp(-j\omega\tau_2) \qquad (7.26)$$

Notice that if $s(t)$ is synchronized with the receiver reference code, the delayed signal, $s(t - \tau_2)$, will not be synchronized since it arrives "$\Delta\tau$" time later. Even though the indirect signal is identical to the direct signal, its delayed version will not correlate properly and may even appear as additional noise into the receiver.

The reflected wave(s) may be specular or diffused. The different ray(s) depend on the properties of the reflecting medium, such as absorbability and dielectric constant. Some arriving stray signals will be diffused and some will be specular in nature. If no single component dominates, the signals will have Rayleigh statistics. A strong component will, among others, manifest Ricean statistics.

Generally, the reflected signals are weaker than the direct wave. However, they are highly specular and can have a component directed at the receiver which can be larger than the direct wave. The receiver, in most cases, favors locking onto this wave.

7.6.1 RAKEing in Performance

Generally, signals arriving at the input of a spread spectrum receiver will arrive via several paths. If the differential delay of the signals is separated by greater than the chip duration, they do not cause interference because they are not synchronized and are not correlated with the local despreader code (for example, see Figure 7–11).

Assume the strongest multipath signal is the desired signal and enters the receiver first. This signal is processed, and out comes the desired data. A second multipath signal, which is delayed by one chip, is independent and is not processed by the receiver as a bona fide signal since it is not correlated (not in sync) with the receiver's built-in code. For example, in the QUALCOMM cellular IS-95 (CDMA), the chip rate is 1.25 chips/second. The chip duration is 0.81 μsec, which corresponds to roughly 240 meters in time. Since an electromagnetic wave travels at roughly one foot in one nanosecond, a

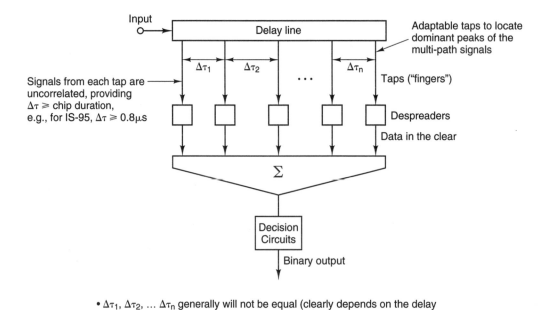

Figure 7–11 Time diversity using a RAKE receiver.

multipath signal delayed by greater than 240 meters is greatly attenuated but does add some noise (hash).

Delayed signals which are not in sync with the local code can be adaptively synchronized with the resident code and thus add to the performance of the receiver. This concept can be implemented with a RAKE receiver [11,12,13,14], which aligns selected multipath signals (strongest selected) with the receiver's despreading code. The RAKE receiver can separately receive, process, and combine the multiple signals. The despread multipath signals are algebraically combined and contribute to the output. It is interesting to note that combining signals which arrive at different times at the receiver is a form of time diversity.

7.6.2 RAKE Receiver

A simplified version of a RAKE receiver is depicted in Figure 7–11. In the version shown, the input multipath signals are passed into a tap delay line with differential delays corresponding to different multipath signal delays. Each tap, which is adaptively located, culls out the strongest multipath signals for processing. In a typical spread spectrum system used in cellular, the number of taps are limited to three or four, and these are often referred to as "fingers".

For a RAKE receiver to be effective in combining multipath signals, the signals must be uncorrelated and separated in time greater than the chip width duration, that is, the multipath signals processed must arrive with a delay spread separation greater than the chip duration. If the delay between consecutive paths is less than a chip duration, the two path signals will coalesce and appear as one. For IS-95 (CDMA), the chip width is 0.8 µs. Therefore, multipath components can be effectively added if tap differential delays are $\Delta\tau \geq 0.8$ µs. This corresponds to a multipath separation of 0.24 km (0.15 mi). It is clear that as the chip duration becomes longer, the number of multipath peaks which can be resolved become fewer.

7.7 Purely Random or Pseudo-Random—What's the Difference?

If a sequence of 1s and 0s is generated by a shift register, a deterministic device, this is not a random process, but is designated a pseudo-random sequence. However, if the 1s and 0s are produced by the flip of a coin, this is purely random because we do not know apriori what the next value will be.

Random codes are not useful in spread spectrum since codes must be known to both the transmitter and receiver.

7.8 Conclusions

As in any system, CDMA offers some advantages over other communication systems. However, it also has some disadvantages. Both are summarized below.

Advantages:
- Low power densities.
- Signals appear noise-like.
- Low detectability due to low power densities; therefore, low probability of intercept (LPI).
- Some measure of message privacy unless adversary has replica of spreading code.
- Multipath tolerance capability.
- Processing gain permits operation in presence of low signal-to-interference ratios.
- can operate with no central timing source.

Disadvantages:

- Requires additional spectrum.
- Needs power control if near-far problem is to be mitigated.
- Hash-limited after a number of accesses.
- Requires synchronization to recover desired signal.
- Requires power control if uplinked to satellite (to prevent power robbing).

7.9 Glossary of Terms

Autocorrelation: Degree of similarity between a sequence and a time shifted replica of itself. Pseudo-random sequences have a strong correlation at zero shift point and weak correlation for shifts other than zero.

Chip: In a spread spectrum system, a chip is a simple element of the spreading code, that is, the duration of one pulse.

Chip rate: The number of chips (pulses) or code elements per second generated by the spreading code generator.

Clock: Clock pulses synchronize the timing of the shift register used in generating a spread spectrum code sequence. A shift register is an electronic circuit of delay elements which can store one binary digit each. These delay elements, called shift register stages, are connected in series. When these stages receive a clock pulse, the binary digit stored in each stage is shifted forward to the next stage. A simple shift register is shown below. Each stage is a flip-flop circuit.

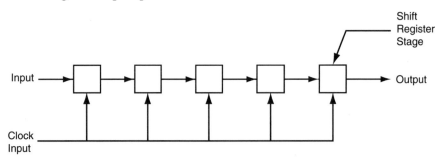

Code Division Multiple Access (CDMA): Several spread spectrum systems accessing a transponder in a satellite (for example) simultaneously.

Code Division Multiplex: A spread spectrum system in which each signal uses a unique spreading code. These codes should be orthogonal to each other to avoid mutual interference.

Cross-correlation: Degree of similarity between two different sequences. Two different orthogonal sequences have zero correlation between them. The fewer the prime factors in the sequence, the less the cross-correlation. For example, for $n = 5$, sequence length is 31 and there is only one prime factor. In this case n has been referred to as a primitive, or Mersenne prime.

Gold Codes: Codes manifesting bounded cross-correlation functions. Can be generated by modulo-2 addition of two preferred maximal-length sequences, but they themselves generally are not maximal-length sequences. Used in CDMA systems since they are quasi-orthogonal, but bounded. Generally maximal-length sequences are not suitable for CDMA.

Interference margin: The amount of interference a system is capable of withstanding while still providing necessary S/N ratio for reliable reception. This is given as

$$M_I = PG - L - (S/N)_{req}$$

where PG is the processing gain, L is the implementation loss, and $(S/N)_{req}$ is the required output S/N ratio after demodulation (this is degraded by interference). The larger the ratio of the BW of the spread signal to that of the information signal, the smaller the effect of unwanted signal interference.

Low probability of intercept (LIP): A term indicating that a signal is difficult to detect. Spread spectrum signals fall into this category. For spread spectrum, a signal may be deep in noise.

Maximal-length sequence: Longest sequence that can be generated by a given length, n, stage shift register.

Mersenne prime: Linear maximal-length sequences that have code lengths equal to a prime number. Example: 3,7,31,127, ... (the fewer the prime factors in a sequence, the smaller the cross-correlation values).

Modulo-2 addition: Arithmetic addition of binary digits. Examples of binary addition are:

$$0 \oplus 0 = 0$$
$$0 \oplus 1 = 1$$
$$1 \oplus 0 = 1$$
$$1 \oplus 1 = 0$$

Near-far performance: A nonrelated user which is close to a receiver will interfere with a bona fide signal which is matched to that receiver, but farther away. This problem is alleviated by automatic power control.

Orthogonal: When two sequences correlated (cross-correlated) with respect to each other are zero, the sequences are said to be orthogonal. Codes that have cross-correlation coefficients (number of bit agreements minus bit disagreements) equal to zero for all pairs in the set of codes are also orthogonal.

e.g., code 1 : 101101
      2 : 110001

The first, fifth, and sixth bits are in agreement = 3
The second, third, and fourth are in disagreement = 3

Therefore, since the number of agreements − number of disagreements = 0, the codes are orthogonal and their cross-correlation is equal to zero.

Processing gain: In spread spectrum, the ratio of the RF bandwidth of the spread signal to the information bit rate.

$$PG = 10 \log_{10}(BW_{RF}/\text{data rate}) \text{ dB}$$

Pseudo-noise (PN): A signal that appears noise-like, but is generated by deterministic means (shift register) and is repetitive.

Random sequence: A sequence that is completely random; an unpredictable sequence (e.g., flip of an unbiased coin).

Shift register generator (SRG): A sequence generator that uses a shift register and modulo-2 adder to generate a pseudo-random sequence.

Spreading code: Binary sequence used to spread the spectrum of direct sequence signals. Its rate is several times greater than the baseband (intelligence) signal.

7.10 References

[1] R. Gold, "Optimal Binary Sequences for Spread Spectrum Multiplexing," *IEEE Trans, on Information Theory*, October 1967.

[2] S.W. Golomb, et al., *Digital Communications, With Space Applications*, Prentice Hall, Englewood Cliffs, NJ, 1957.

[3] R. Moser and J. Stover, "Generation of Pseudo-Random Sequences for Spread Spectrum Systems," *Microwave Journal*, May 1985.

[4] R.C. Dixon, *Spread Spectrum Systems*, Wiley & Sons, New York, 1976.

[5] NSA, *Spread Spectrum Handbook*, March 1979.

[6] R.H. Petit, *ECM and ECCM Techniques for Digital Communications System*, Lifetime Learning Publications, 1982.

[7] R. Gold, "Study of Correlation Properties of Binary Sequences," Magnavox Interim Report #1, January 1964.

[8] W.J. Judge, "Multiplexing Using Quasiorthogonal Binary Functions," *AIEE Trans. Communications Electronics*, May 1962.

[9] B. Pattan, "A Tutorial Look at Spread Spectrum," FCC OET Internal Memorandum, April 23, 1990.

[10] F.J. Mac Williams and N.J. Slaone, "Pseudo Random Sequences and Arrays," *Proc. IEEE*, December 1976.

[11] R. Price and P.E. Green, "A Communication Technique for Multipath Channels," *Proc. IRE*, March 1958.

[12] U. Grab, et al., "Microcellular Direct-Sequence Spread-Spectrum Radio System Using N-Path RAKE Receiver," *IEEE J. Select. Areas Comm.*, June 1990.

[13] D.R. Bitzer, et al., "A Rake System for Tropospheric Scatter," *IEEE Trans. on Comm. Tech.*, August 1966.

[14] W.W. Ward, "The NONAC and RAKE Systems," *The Lincoln Laboratory Journal*, Vol. 5, No. 3, 1992.

CHAPTER 8

Terrestrial-based Wireless Communications

8.1 Introduction

Cellular mobile telephony is now in its second decade of unprecedented growth. Its early development was performed at AT&T in the late 1970s, and validation tests were performed in Chicago and Newark, N.J. The first commercial cellular service in the U.S. began in Chicago in October 1983. The evolution began at the single cell concept, where a single transmitter/receiver communicated with a mobile within range of the cell's signal limit, and the range was dependent on the transmitter EIRP (Effective Isotropic Radiated Power) and mobile EIRP. It is clear that any spectrum assigned to this service was limited to the number of channels that could be fitted into the allocated spectral band. To prevent inter-channel interference, no co-channel operation was possible. For example, if the bandwidth of the spectrum available was 25 MHz and the channel bandwidth was 30 MHz, the theoretical number of conversations in this big cell concept was 25 MHz / 30 kHz = 833. This was therefore the maximum number of users in the available spectrum segment. This figure is relatively small, especially in urban areas, and we might ask what does one do to provide capacity for thousands of users in the same general area?

The solution proposed and implemented was to divide a service or market area into a cluster of idealistic tessellated hexagonal cells, where only a subset of the channels could be reused in cells which were sufficiently separated. The separation criterion was the carrier-to-noise ratio at the receiver that was acceptable to the user.

Actually the separation that allows co-channel acceptability is shorter than that given by the $1/R_2$ relationship (R = line-of-sight distance), since the propagation anomalies between the transmitter (base station) within the cell and receiver (mobile) are much more attentuative due to multipath and shadowing losses, and generally the path is rarely line-of-sight. Additionally, propagation anomalies are one of the most difficult problems to quantify in cellular systems.

Each cell site contains a transmitter/receiver to communicate with a mobile receiver within its domain. The fixed transmitter/receiver is referred to as the base station, or cell site.

It is customary to depict cells as hexagonal in shape since this approximates the omni-directional coverage of a cell site antenna and further leaves no gaps when tessellated.[1] In actuality, the real-world cell boundaries are more amorphous and a cell cluster may look more like a jigsaw puzzle. A real-world representative signal profile in a virtual cell is depicted in Figure 8–1. Within the cell there may be drop-out areas and isolated pockets. The biggest contributors to these anomalies are multipath, shadowing, and blockage.

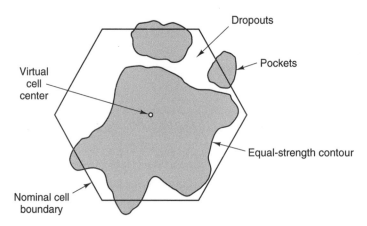

- Grey areas define signal levels greater than the criterion for acceptable service.
- For any practical system, topology and limitations in the choice of the cell sites can make the performance very different from that estimated under ideal conditions.

Figure 8–1 A representative signal profile in a hypothetical cell boundary.

---

1. Juxtaposing cells with continuous coverage (only squares, equilateral triangles, and hexagons can be tessellated).

Introduction

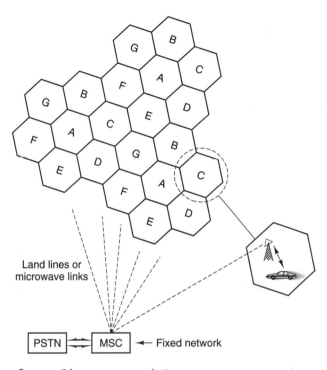

- Seven-cell frequency reuse cluster.

Figure 8-2 Idealistic cellular structure. Practically the cellular contours are more amorphously shaped, depending on propagation vagaries.

Clearly, the base station tower location and height has some influence, as well as the mobile's movement within the cell. Transmissions from the base station to the mobile (downlink) and from the mobile to the base station (uplink) are subjected to various perturbations. In fact, the statistics of the forward transmissions (downlink) and reverse transmissions (uplink) are different. They see different anomalies, in addition to transmitting in different bands, i.e., on different frequencies.

A typical cellular topology which was proposed during the early stages and is still in use today is the seven-cell, tessellated cell cluster. This is shown in Figure 8-2. For American cellular systems, the cell radii may range from 2–20 km. Each cell includes a base station (cell site) which communicates (forward link) with the mobile within the cell as shown in Figure 8-3. The base station antenna may be centered in the cell, offering an omni-directional antenna pattern in azimuth with sufficient EIRP to serve the extent of the cell. Another base station arrangement is to deploy the tower at a hexago-

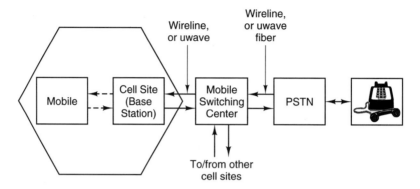

Figure 8–3 Connectivity for wireless cellular communications.

nal vertex, with directional beams serving three cells. The directional beams have beamwidths of nominally 120 degrees. The advantage of using directional beams is that the overall interference to a receiver is reduced because some beams in other cells are not illuminated by a directional beam. Sectorization is more expensive than using centered omni-directional antennas, but it can serve a greater number of cells with fewer antennas than if omnis were used in each cell. It therefore increases the capacity of the system.

The base stations communicate with the Mobile Switching Center (MSC), which interfaces with the local PSTN, or land-line network. The MSC monitors the movement of mobile units and automatically switches mobile transmissions as the mobile station moves from one cell to another. There is a lot of computer power at the MSC. Intercell movement is referred to as handoff, or handover. Since contiguous cells are operating at different frequencies (or channels) for present American cellular systems (to prevent interference), the carrier frequencies must change at the interface. This applies to both analog and digital systems in operation today. In third-generation digital cellular systems (CDMA, IS-95), interference is controlled by using orthogonal coding as opposed to frequency change between the contiguous cells. Here, cells operate at the same frequency, which reduces the burden at handoff.

8.2 Frequency Bands of Operation

The high end of the UHF band is allocated to mobile cellular operation in the U.S. This is indicated in Figure 8–4. The band 869-894 MHz (downlink) is used for transmissions from the base station in a cell to the mobile within the cell. The band 824-849

Frequency Bands of Operation

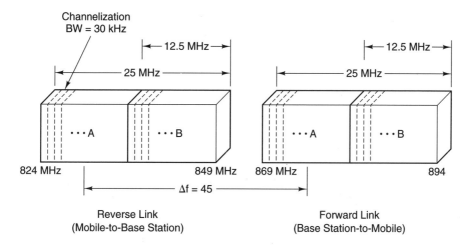

Figure 8–4 Cellular allocations in the U.S.

MHz (uplink) is used from the mobile to the base station (cell site). The bandwidth of each is 25 MHz, and the two bands are separated by 45 MHz.

The first cellular system developed in the world used FM modulation. This was the so-called Advanced Mobile Phone System (AMPS). The maximum frequency deviation of the AMPS signal was $f = 12$ kHz. The channel transmission bandwidth (from Carson's Rule) was therefore

$$B = 2\Delta f + 2W = 2(12) + 2(3 \text{ kHz}) = 30 \text{ kHz} \tag{8.1}$$

For operation above threshold, the output was

$$S/N \approx (3/2)\beta^2 (S/N) \text{ in, where } \beta = \Delta f/W \tag{8.2}$$

As the modulation format suggests, the allocated spectrum was broken up into several channels with users assigned to discrete channels. For the AMPS system, the channel width or carrier spacing was 30 kHz, as indicated.

Cellular capacity is generally given by the number of simultaneous users, S, per base station of a cell site. For a given amount of spectrum (forward or reverse), the maximum number of users (including some saved channels for signaling and control) is therefore

$$S = (1/N)(B_a/B_v) \tag{8.3}$$

where B_v is the bandwidth of the voice channel, B_a is allocated spectrum, and N is the reuse factor and a function of system resistance to co-channel interference. For the AMPS system, due to the hexagonal cell format, a seven-cell cluster was commonly used. Therefore,

$$S = (1/7)\ (25\ \text{MHz}/30\ \text{kHz}) = 120\ \text{channels in each cell} \qquad (8.4)$$

For the new-generation cellular systems using digital technology (GSM/TDMA, TDMA, CDMA), the number of channels per cell is at least three to ten times as much. This comment will be explained below.

The FCC partitions the spectrum between two operators in a given market area. For example, for the Washington area, Cellular One and Bell Atlantic use 12.5 MHz (in each direction) in each market area for a seven-cell topology. In other words, each cell uses 12.5 MHz / 7, or 1.18 MHz of bandwidth, which is divided up among the channels in a cell. Therefore, for the AMPS system in which the channels are 30 kHz wide, the number of channels per cell is *60 for each operator.* As indicated previously, some channels are assigned for signaling and control. These channels will use digital modulation (FSK) in lieu of analog for this function. In the American digital cellular system (IS-54), there are three users per channel (in 30 kHz), $3 \times 60 = 180$, or a threefold increase in capacity.

The areas beyond a unit's seven-cell cluster are repeated over and over again until the full service area is blanketed. Contiguous clusters use the same frequency set, and each subset in a cluster is identical to the same subset in the contiguous cluster. As indicated previously, identical subsets in separate clusters must be sufficiently separated to reduce mutual interference. The criterion which has been established for analog AMPS is the carrier-to-interference ratio, *C/I*, which should be 17–18 dB for the seven-cell pattern. A digital modulation system is more robust and can operate with a lower *C/I*. For example, for the European GSM system, the acceptable *C/I* is 12 dB. This can therefore be reflected as a smaller cell structure (still macro), or a four-cell (in lieu of seven-cell) repeat pattern. In other words, the lower the signal-to-interference ratio a receiver can work with, the shorter the reuse distance with which co-channel cells can repeat. We thus have greater capacity.

8.3 Interference Analysis

For a seven-cell cluster, it is noticed from the ensemble of clusters shown in Figure 8–5, that the totality of interference stems from six "number 1" cells, which all operate on the same frequency. The base station antennas are assumed to be located at the cell center, and omni-directional antenna patterns are used.

Interference Analysis

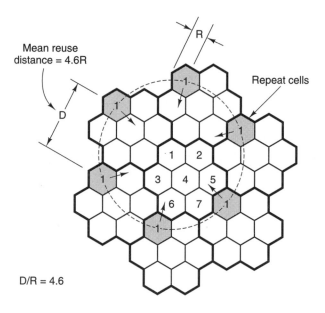

- D/R: co-channel reuse ratio.
- D: separation between two co-channel cells.
- R: radius of cells.
- In hex-shaped cellular mosaic, D/R = $\sqrt{3N}$.
- N: number of cells in cluster.

Figure 8–5 Ideal seven-cell cluster repeat pattern geometry.

These cells must be separated sufficiently to prevent excessive interference. The ratio of cell separation to cell radius is referred to as the *co-channel reuse ratio*, *D/R*. For the hexagonal-shaped cellular system, this ratio is equal to

$$D/R = \sqrt{3N} \qquad (8.5)$$

where *N* is cluster cellular size [2]. The *C/I* ratio for the topology shown in Figure 8–5 is simply

$$C/I = C/\sum_{1}^{6} I \qquad (8.6)$$

If a mobile is located at the edge of a cell at distance *r*, the interference is at a distance *D*, and all signals (bona fide + interference) are equal, we have the following signals at the bona fide receiver (center)

$$C/I = (k/R^4)/(6k/D^4) = D^4/6R^4 \qquad (8.7)$$

where we have assumed the *Hata model*, which is where the signals fall off as one over the distance to the fourth power (40 dB/decade). Bell Laboratories has also made this observation, which we alluded to previously. Signals in cellular systems fall off at a rate greater than the free-space loss of one over the distance squared. If we plug Equation (8.5) into Equation (8.7) we obtain

$$C/I = (1/6)[\sqrt{3N}]^4 = (1/6)[9N^2] = 1.5N^2 \tag{8.8}$$

And, since there are six peripheral interferers in the cell arrangement (Figure 8–5), we get

$$C/I = 1.5(6)^2 = 54$$
$$\text{or} \quad C/I = 17.3 \text{ dB} \tag{8.9}$$

There may be another mitigating phenomenon which comes into play here. In a situation in which two signals are at the same frequency and where one is weaker than the other, the stronger signal will suppress the weaker or interfering signal because of the capture effect of FM (AMPS uses FM). However, the *D/R* factor is still used to prevent a spurious signal from capturing the bona fide signal.

For a seven-cell cluster system, the co-channel reuse ratio is thus

$$D/R = \sqrt{3N} = \sqrt{21} = 4.6 \tag{8.10}$$

For a cell with a radius of 3 km (1.8 miles), for example, the distance between cell centers using the same frequency must be 13 km, or 7.8 miles.

We also observe from Equation (8.5) that if the number of cells (*N*) in the cluster is increased, we can increase the distance before frequency repeat occurs and therefore reduce co-channel interference. However, the capacity would drop since the total number of channels (dictated by the amount of spectrum allocated) in the larger cluster would be divided by the number of cells in the expanded cluster. Therefore, the capacity per cell would be reduced.

Notice that the separation between cells to avoid co-channel interference is not specifically related to actual miles, but the ratio of *D/R*, which is equal to 4.6. So, whether we start with large cells or small cells, as long as this ratio is maintained, we will be limited by the minimum acceptable signal-to-interference ratio, 18 dB.

In Figure 8–5 we assumed that the base stations were located at the center of the cells. This implies that the cell site transceivers use omni-directional antennas. In more updated systems, directional antennas are used and located at the vertex of a cell, as shown in Figure 8–6. At a cell site, three directional antennas are used, with each serving a different cell. The ideal beamwidth for each 120° sector theoretically assumes no

Interference Analysis

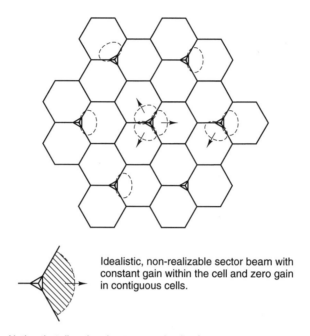

Figure 8–6 Cellular structure using directional antennas at the base stations.

radiation outside the parent cell. Readers versed in antenna technology should realize that *idealized* sector beams are difficult to achieve, and the azimuthal pattern spills over into adjacent sectors of the cell.

In addition, it is desirable and realizable to have directionality in the vertical direction, which further increases the overall antenna gain. This also affords some protection in reducing multipath interference to receiving mobile stations. There is also flexibility in tilting the vertical pattern to improve performance. This artifice is used in practice.

The use of directional antennas substantially reduces the radiation to mobiles, which are not in the main beam. For a seven-cell cluster system, there is a better than 4-dB improvement in *C/I*. Looking at it differently, this reduces the co-channel reuse factor, or the spacing between co-channel cells. This corresponds to a smaller number of cells per cluster, or a smaller number of channel sets. Since the total number of channels is fixed (allocated spectrum/channel bandwidth), the smaller number of sets means more channels per set and per cell site. For example, for AMPS, we have allocated spectrum/

channel BW = 25 MHz / 30 kHz = 840 channels fixed. If N is reduced from 7 to 4 (channel sets), because of the benefit derived from the use of directional beams, we have 840 / 4 = 210 number of channels served by a directional base station. Therefore, each cell (sector) in the triad will receive 70 channels. You would think that this would be a disadvantage since omni cells can serve 120 users, which reduces the trunking efficiency (with increased blocking). However, capacity is enhanced by reduced cluster size.

The chief motivation in using directional antennas at cell sites is to reduce the cost since fewer are required than omni cell sites which use an antenna in each cell. Real estate for cell sites is expensive and has caused continued zoning problems in many localities (see comments at the end of this chapter).

The center-fed cells are more likely to be used in rural areas, or small cities which are start-up systems. These areas may later scale up to sectorization as capacity warrants. Also, omnis (monopoles) will probably prevail in PCS microcells where there is a high density of cells.

The interference scenario for the IS-95 standard using CDMA (spread spectrum) differs from that for the analog-based AMPS system. Separation between users in FDMA and TDMA systems is achieved by the use of different frequency bins and time slots, respectively. In CDMA, however, separation is achieved via the use of different codes imparted to the signal, and each cell uses the same frequency. The codes are orthogonal (almost) to each other and thus ideally provide zero correlation between each other. As discussed in the previous chapter, the most popular coding schemes are direct sequences using pseudo-random codes to modulate (modulo-2 addition) the data bits. This increases the bandwidth of the transmitted signal. Coded signals are not actually orthogonal to each other and thus there is some correlation between the unwanted coded signal and the receiver's built-in code, which occurs in the despreading process, and which appears to be noise-like, thus adding to the receiver thermal noise. This correlation noise (frequently referred to as hash) in effect increases the receiver noise figure, or reduces the E_b / N_o' where N_o' contains both receiver thermal noise and correlation noise. The cellular receivers will therefore be perturbed by other coded waveforms which are not matched to the receiver's code.

Even though cellular communication systems use the same frequency in neighboring cells, the reuse efficiency is not equal to one. Since cells peripheral to the one in question will contribute unmatched coded noise (hash) to the surrounded cell (see Figure 8–7), the neighboring cells contribute interference, which is about half of the hash noise contributed by the *self*-generated correlation noise in the victim cell. The reuse efficiency for an omni cell is 1 / 1.5 = 0.67. As will be shown in the following, this will have some degrading effect on system capacity.

Interference Analysis

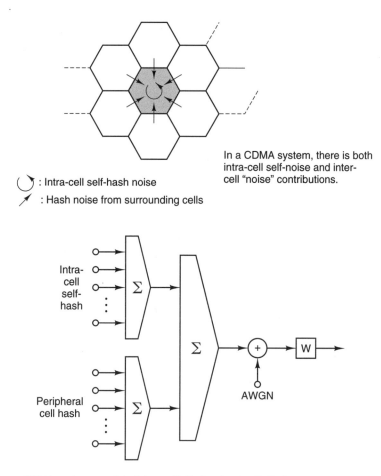

- CDMA systems are interference-limited, not noise-limited.

Figure 8–7 Correlation noise (hash) limits the capacity of a CDMA system.

In a seven-cell cluster geometry, the six cells surrounding the bona fide cell will contribute correlation noise, which adds to the bona fide receiver's thermal noise. In addition, as previously stated, channels within the bona fide cell will also contribute their self-noise. The self-noise is the energy contribution within the cell, given as $(M-1)$Echip, where M is the total number of channels in the cell. Note that each signal emanating from the base station in the cell is attenuated by the $1/R$ factor. In addition, the peripheral cells [6] are adding their contribution with less contribution from the next tier of cells (see Figure 8–7).

The undesired CDMA signals appear as noise at the bona fide receiver output. The equivalent E_b/N_o' is given by

$$E_b / N_o' = (E_b = P / Rb) / [N_o + (M - 1) E_{\text{chip}})] \tag{8.11}$$

where M is total number of CDMA users in proximity to the victim receiver and Echip is energy per chip for $M - 1$ interfering chips, and statistics similar to N_o. Its noise power is $N = (N_o' - N_o)W$, (W = bandwidth).

The above assumes equal power in received signals. This is a valid assumption. Also note that the second term in the denominator of the right member is the self-noise due to uncorrelated noise (hash) within a cell, *plus* uncorrelated noise from surrounding cells. For example, consider the six cells in a seven-cell cluster. This noise is illuminating a receiver in the central cell. There are also contributions from cells outside the seven-cell cluster, but these are less significant due to the range falloff factor of $1 / R^4$.

The uncorrelated noise term in the above expression can be broken down into the self-noise (within the center cell) and the peripheral cells' uncorrelated noise contribution. Note that the self-noise from the center cell (where the bona fide receiver or victim receiver is located) plus the noise of the surrounding cells is range dependent. We therefore have the following:

$$E_b / N_o' = \frac{E_b}{N_o + \dfrac{(M_s - 1)E_{\text{chip}}}{R_s^4} + \dfrac{6M_p(E_{\text{chip}})}{R_p^4}} \tag{8.12}$$

where R_s is the range from the base station transmitting M different coded signals within the center cell, R_p is the range from the peripheral cells' (6) transmitters, assuming equal power, and E_{chip}, or energy per chip, $= PT_c = P / R_c$, where T_c = chip width. E_b is $PT_b = P / R_b$, where T_b = data bit width (at receiver), N_o is receiver thermal noise, and M_p is non-matched CDMA signals from the six peripheral cells illuminating the receiver in the central cell. It has been determined that the interference contribution from neighboring cells is half of that of sources within a cell. This will be shown below to have an impact on the capacity of a cell.

If we divide the numerator and denominator in the right member by E_b, and realizing that $E_c = PT_c$ and $E_b = PT_b$, we obtain

$$(E_b / N_o') = \frac{1}{(N_o / E_b) + \dfrac{(M_s - 1)PT_c}{PT_b R_s^4} + \dfrac{M_p PT_c}{PT_b R_p^4}} \tag{8.13}$$

All powers are equal, as in

$$\therefore (E_b / N_o') = \frac{1}{(N_o / E_b) + \dfrac{(M_s - 1)T_c}{T_b R_s^4} + \dfrac{M_p T_c}{T_b R_p^4}} \tag{8.14}$$

Interference Analysis

but T_c / T_b = reciprocal of the processing gain. Finally, we get

$$(E_b/N_o') = \frac{1}{(N_o/E_b) + \frac{(M_s - 1)}{\overline{PGR_s^4}} + \frac{M_p}{\overline{PGR_p^4}}} \tag{8.15}$$

and for $PG \gg 1$, we obtain $E_b / N_o' = E_n / N_o$, a benign environment.

If the system requires a threshold value of E_b / N_o to realize a certain BER probability, $p(e)$, uncorrelated noise will have a deleterious effect, and may cause the system to become inoperable if too high. This therefore limits the number of users which can be accommodated. More will be said on this in the next section.

We may also use Equation (8.11) to find how many users can be supported by the bandwidth, W, where $1 / W = R_c$, the chip rate. Assume for a moment we are dealing with a *single* cell with several spread spectrum signals (M) operating within the cell. To repeat Equation (8.11) from above, the equivalent E_b / N_o resulting from the self-noise energy from $M - 1$ transmitters is

$$E_b/N_o' = \frac{E_b}{N_o + (M_s - 1)E_{\text{chip}}} \tag{8.16}$$

where E_{chip} is PT_c energy from an interfering signal, assuming they all have equal power at the bona fide receiver. Assuming N_o is small in comparison to the perturbing noise, we have

$$E_b/N_o' = E_b/(M-1)E_c = (P/R_b)/[(M-1)P/R_c] \tag{8.17}$$

and since all symbols have equal power, we have

$$E_b/N_o' = [1/(M-1)](R_c/R_b) \tag{8.18}$$

where R_c / R_b = chip rate / data rate = processing gain. (For QUALCOMM's CDMA cellular system, the spreading ratio is 128:1 or 21 dB relative to a 9.6-kbps voice coding rate. The chip width is about 0.81 μsec.)

Assuming a chip rate of $R_c = 1.25$ Mbps (using BPSK), a bit rate of $R_b = 9.6$ kbps, and a required $E_b / N_o = 7$ dB to realize $p(e) = 10^{-4}$, we obtain

$$E_b/N_o' \text{ dB} = 10\log(R_c/R_b) - 10\log(M-1) \tag{8.19}$$

$$10\log(M-1) = 10\log(R_c/R_b) - (E_b/N_o') \tag{8.20}$$

$$\therefore M = 14 \text{ users} \tag{8.21}$$

Incorporating the voice activation factor of 2, sectorization gain of 3, and a factor for effective interference from neighboring cells relative to interference from its own cell, this factor is about 0.67 [5]. Although the actual frequency reuse of CDMA is equal to one, the effective frequency reuse is less than one due to the interference from users in the other cells. The capacity is thus reduced by the factor $1/1.5 = 0.67$. Therefore, the number of users per cell is

$$M' = 14\,(2)\,(3)\,(0.67) = 56 \text{ users per cell} / 1.25 \text{ MHz} \qquad (8.22)$$

The capacity of an AMPS system is six subscribers per cell per 1.25 MHz (840 / 7 = 120 channels/cell/25MHz). Therefore, theoretically a CDMA system has approximately nine times the capacity of an AMPS system for a comparable bandwidth (1.25 MHz). I have seen early figures like 10 to 20 times better than AMPS, but I am not sure all investigators included self-noise in their calculations. A number like 20 is idealistic. Some of the latest studies and tests seem to indicate the increase in capacity may even be less than 10.

8.4 Increasing Capacity

Prior to cellular mobile radios, all systems were single-cell. During the early years of cellular radio, it was soon realized that with the amount of spectrum allocated to cellular, one big cell providing 833 FM analog channels would not satisfy the anticipated number of users—especially in urban areas. Various expedients were considered to alleviate this shortfall in capacity. One proposal was to partition a service area into contiguous, macroscopic-sized, cellular, hexagonal structures, with only a subset of radio channels being used within a cell. There would be reuse of the frequencies in cells that were sufficiently separated. The cell sizes would vary from 10–20 miles in radius.

As the desire for cellular phones burgeoned, capacity was approaching saturation. This was somewhat alleviated by *cell splitting*, where the areas having the highest traffic density used smaller cells. The smaller cells had the same cell capacity, therefore accommodating a higher density of subscribers. These were referred to as micro cells with radii down to one mile. We discussed this concept in previous sections.

Figure 8–8 shows a macro cell going through various evolutions of cell splitting [3]. At each split, the nascent cell is one-half the diameter of the previous cell. It has been shown that a 50% reduction in cell radius should yield a fourfold increase in traffic capacity. With each split, there is an increase in capacity. Ultimately, the last split indicates cells which are about one mile in radius. Naturally there is frequency reuse in the last display, and each cell can support the same number of users as the macro cell, but the density of users has increased.

Increasing Capacity

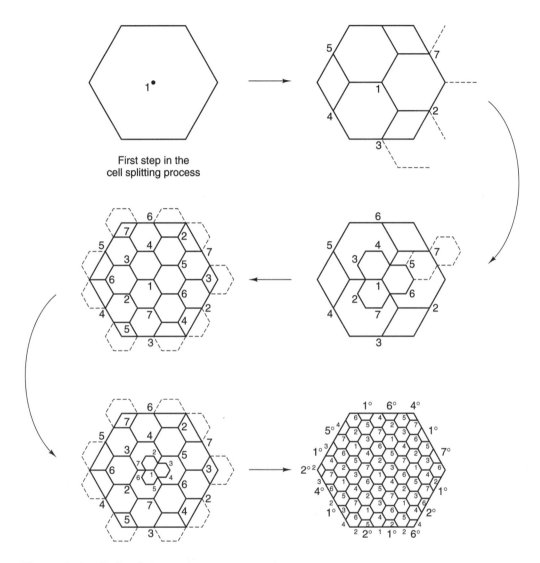

Figure 8-8 Cell splitting to increase capacity.

Figure 8–9 shows a service area which was initially blanketed by macro cells, and as the traffic density grew, it was partitioned into micro cells. Micro cells have higher capacity because the spectrum can be reused many times more than if a macro cell occupied the area. We have increased the number of users per megahertz per square mile, and have thus increased the spectral efficiency by reducing the value of D, the cell reuse separation. The number of channels per cell is determined by the size of the cluster, not the cell size. The capacity of the cell is independent of cell size. The power of the cell is also lower than for the macro cells. This eases the burden on the handset since distances covered are reduced and signal statistics are somewhat more docile. There may even be line-of-sight in some cases since the cell sites may be located along roadways.

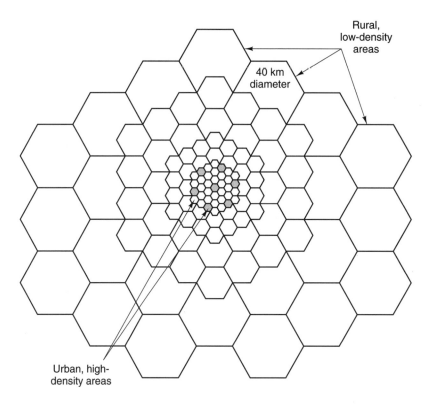

Figure 8–9 Evolution of cells from macro cells to micro cells as traffic density increased, with no additional frequency allocation.

Since traffic density in outlying areas is moderate, these locales continue to be served by macro cells. Notice that center city has the smallest cells, but if the next tier cells are in suburbia, these too could be split as they become more urbanized.

Going to micro cells also brings along concomitant problems. Many more base stations are required, the number of cell-to-cell handovers increases, there are additional costs for base stations and for links back to the MSC, and there is the base station real estate requirement. In the last problem, political considerations take front stage, not technical problems. Everyone "loves" cellular phones, but not base stations in proximity to their home. Of course, the positive side of using micro cells is that no additional frequency allocation is required.

Even though we have considered that the concept of cell splitting increases capacity, this may not be the route to take initially because of the problems alluded to above. One of the biggest advances made to alleviate the capacity shortfall is to go to digital modulation. No additional cell sites are required, but the capacity can be increased by at least three times the capacity of analog modulation (e.g., AMPS). Another concept which does not require additional cell sites is the use of sectorization. That is, in lieu of using an omni antenna pattern in a cell, directional antennas that illuminate three cells from one location are used. This was discussed previously.

Allocating more spectrum to increase capacity is not a feasible option in the long run. Other means must be found in the ever-present spectrum constraint. Innovative technology is the answer. Several areas which have reached some degree of maturity or are being investigated are listed in Table 8–1. The various schemes manifest varying degrees of complexity, performance, and cost. Several listed possibilities have been discussed previously.

Table 8–1 Technology Maturation to Increase System Capacity

Technology	Description of Benefit
Frequency Reuse	At the present time, reuse factors of 1/7 are most popular, but will drop to 1/4 or 1/3 as digital modulation becomes more prevalent; factors close to one are now possible for SSMA/CDMA. Ideally, for the IS-95 DS-CDMA system, we would expect a reuse factor of one since all cells use the same frequency; however, the non-perfect orthogonality of the codes used permit the code residues to enter a non-matched receiver which is noise-like and adds to the receiver's thermal noise. As indicated previously, the surrounding cells contribute about 50% of the correlation noise provided by the parent cell. Therefore, the reuse factor is is 1/1.5 (clearly not one).
Antenna Beam Sectorization	Used in lieu of omni radiation (the latter is employed in starter systems or rural areas); some loss in trunking efficiency because of the reduced number of channels per cell; sectorization provides better than a 4-dB improvement in C/I (in a seven-cell cluster).
Cell Splitting	Going from a macro cell to a micro cell cluster.
Variable Speed Codecs	Transmits at lower rates and reduced power levels when voice activity is low or absent.

(continued)

Table 8–1 Technology Maturation to Increase System Capacity

Technology	Description of Benefit
Reduced Channel Bandwidth	Concomitant reduction in signal bandwidth.
Optimum Modulation	Using digital modulation with a waveform of high spectral efficiency (e.g., GMSK* used to minimize inter-channel interference and thus achieve greater signal packing). Operating companies will selectively convert analog channels to digital operation to relieve traffic congestion at cellular base stations.
Use of Error Correction	FEC + interleaving; FEC reduces random-type errors and interleaving reduces bursty errors due to multi-path.
Antenna Space Diversity	Not practical at handset, but at base station.
Time Diversity	Use of the RAKE receiver at both the base station and handset. Note QUALCOMM's CDMA system uses a three-finger RAKE at the hand set. RAKE gathers time-delayed multi-paths and enhances output.
Increase Source Coding Compression	This reduces the bandwidth required for signal representation. However, you should use caution here since signal fidelity may suffer. At low bit rates, we can realize intelligibility, but lose speaker (gender also) recognition. The European GSM system uses a 13-kbps codec, while the U.S. IS-54 standard uses a 13-kps codec, which originally use 8-kbps. Half-rate codecs have been considered for both, which would accomodate more channels per carrier. We can expect near-toll quality voice.
Voice Activation in Digital Systems	Human voice activity cycle is 35%; the rest of the time we are listening. These dead times can be taken advantage of to increase capacity. For example, this is taken advantage of in IS-95 to increase capacity by roughly a factor of two. In a Hughes E-TDMA architecture, Digital Speech Interpolation (DSI) is used to pack six voice streams in six channels (with 40% utilization) into three physical channels (with 80% utilization).
Accepting Higher C/Is	The present AMP-type systems require a C/I = 18 dB for adequate performance. Digital systems like GSM perform well with a C/I = 12 dB.

* This is becoming a very popular modulation waveform. It is now used in European GSM and other parts of the world, and in the American Sprint Spectrum PCS. Three requirements are necessary for a good modulation waveform: 1) low out-of-band radiation so that adjacent channel interference is low, 2) relatively narrow bandwidth to allow good spectrum efficiency (packing) and 3) constant envelope to allow the use of simple and efficient power amplifiers (e.g., Class-C). A minimum shift keying (MSK) modulation with Gaussian pre-filter satisfies all the above three requirements.

8.5 Cellular Standards

Digital cellular standards being used presently or considered for worldwide application are indicated in Table 8–2. The European GSM digital standard is now in use in 16 European nations, as well as in many other parts of the world. The system uses TDMA with very efficient GMSK modulation. Roaming is now possible all over Europe. The American Digital Cellular (ADC) standard is IS-54, which uses TDMA and the same channel spacing as AMPS, but achieves three times the capacity of the AMPS system. Its modulation is $\pi/4$-DQPSK. It is presently being implemented in the U.S. on a selective basis, and complements the AMPS system in high-traffic areas.

Table 8–2 Three Basic Digital Standards which are Operational Today and Providing Greater Capacity than Analog Systems

System	Features
European GSM	• using TDMA
	• using GMSK modulation
North American Digital Cellular (TDMA)	• IS-95 standard TDMA using $\pi/4$-DQPSK modulation
	• also referred to as D-AMPS-IS-54C
	• using TDMA
Japanese Digital Cellular	• using TDMA
	• using $\pi/4$-DQPSK
1. Another U.S. digital standard is IS-95, using DS-CDMA with QPSK modulation.	
2. Operation companies are selectively converting analog channels to digital operation to relieve traffic congestion at cellular base stations. However, AMPS is entrenched, so it will be available for years to come.	

Because of the volume of AMP handsets in use, it is doubtful that the standard will be phased out in the near future. As cellular networks become more of a mixture, dual mode phones will be implemented.

Another digital standard which has been considered by (not U.S. standard) TIA is IS-95, which uses QPSK modulation with CDMA. The QUALCOMM IS-95 system initially uses about 1/10 of the allocated spectrum per operator in a service area, and all users use the same spectrum. Additional demand will open up other nominal 1.25-MHz slots. In this standard, common pseudo random code is used for spreading, and a user is tagged by superimposing Walsh function orthogonal codes. This is considered to be the third-generation cellular system, and early claims have been made that it will have 10 to 20 times the capacity of the AMPS system. The last word has yet to be said on this. Some operators are now implementing CDMA and others are mixing AMPS and CDMA and/or TDMA.

The Japanese have a digital system (JDC) similar to the American IS-54, using $\pi/4$-QPSK with three users per channel. The parameters of several of these digital cellular standards are described in Tables 8–3 through 8–7. Table 8–6 shows the bandwidth efficiency of the various standards. Note the efficiency of IS-54 is about 20% more than GSM.

Table 8–7 lists some of the benefits that may be realized from CDMA. Some of those listed are applicable to other cellular standards as well, but some are unique to CDMA only.

Table 8–3 Cellular Standards Parameters

	IS-54	GSM	CDMA
R_{burst}	48.6 Kbps	270.83 Kbps	1.2288 Mchips/sec
Modulation	$\pi/4$-QPSK	GMSK ($B_b / R_b = .3$)	"QPSK"
R_{symbol}	24.3 Ks/sec	270.83 Ks/s	1.2288 Mchips/sec
T_{symbol}	41.2 μs*	3.69 μs**	.81 μs/chip
Channel Spacing (Δf)	30 KHz	200 KHz	1.25 MHz
Multiple Access	TDMA (3 users)	TDMA (8 users)	CDMA (128 chips/bit)
Bandwidth Efficiency ($R_{burst} / \Delta f$)	1.62 bps/Hz (48.6/30)	1.35 bps/Hz (270.83/200)	.983 chips/sec//Hz (1.2288/1.25)
Spectral Packing ($\Delta f / R_{burst}$)	.62 Hz/bps (30/48.6)	.738 Hz/bps (200/270.83)	1.02 Hz/bps (1.25/1.2288)
Voice Coding	VSELP	LPC-RPE	CELP
$R_{VOCODER}$	7.95 Kpbs	13 Kbps	9.6 Kbps (8.0 Kbps)
R_{bit} / Channel Coding	Rate 1/2 Convolutional + Int	Rate 1/2 Convolutional + Int	Rate 1/2 Convolutional + Int
R_{bit} / Channel (after coding)	13 Kbps		
$R_{bit\ (overall)}$ / Channel	16.2 Kbps	22.8 Kbps	9.6 Kbps

(continued)

Table 8–3 Cellular Standards Parameters (Continued)

	IS-54	GSM	CDMA
Miscellaneous	• Same channel spacing as AMPS. • Same frequency range. • BW efficiency: 1.62 b/s-Hz. • Three signals in 1 AMPS channel. • Roughly the same C/I as AMPS. • Not constant envelope modulation.		• Delay spread >1 μsec is uncorrelated with code (chip length is 0.81 usec).

*   Fading Model - Two Raleigh faded rays with maximum time delay of 41 μs (one symbol only).
** Fading Model - Six fading rays with maximum time delay of 16 μs (four symbol intervals).

Table 8–4 Cellular Access Matrix

	AMPS (IS-3)	E-TDMA	IS-54	GSM	JDCRS	Q-CDMA
Frequency Band Forward Band (MHz) Reverse Band (MHz)	869-894 824-849	869-894 824-849	869-894 824-849	935-960 890-915	810-830 940-980	825.03-834.33 870.03-879.33
Spectrum (MHz)	50	50	50	50	40	9.3
Duplex Channel Bandwidth (kHz)	60 (2×30)	60 (2×30)	60 (2×30)	400 (2×200)	50 (25×2)	2500 (2×1250)
Users per Carrier	1	5-10	3	8	3	60
Max No. Users in Available Spectrum	832	4160-8300	2496 or 4992	1000	2400-4800	480
Coder Bit Rate (kbps)	NA	–4.2	7.95 or 6.1	13	6.7	9.6
Voice Coding Method	NA	VSELP	VSELP	REP-LTP	VSELP	QCELP
Frame Duration (ms)	40	40	40	4.615		20
Duplex Method	FDD	FDD	FDD	FDD	FDD	FDD
Access Method	FDMA	TDMA	TDMA	TDMA	TDMA	CDMA

(continued)

Table 8-4 Cellular Access Matrix

	AMPS (IS-3)	E-TDMA	IS-54	GSM	JDCRS	Q-CDMA
Modulation Scheme	Analog FM	π/4 DQPSK	π/4 DQPSK	GMSK	π/4 DQPSK	QPSK
Handset Transmit Power, Max/Avg	6.3 W (max) 4.0 W Class I 0.6 W Class IV	6.3 W (max) 4.0 W Class I 0.6 W Class IV	6.3 W (max) 4.0 W Class I 0.6 W Class IV	5 classes 20 W Class 1.8 W Class V		6.3 W (max)
Channel Allocation Method	Static	Static	Static	Static		Dynamic
Coordinated Frequency Sharing	No	No	No	No		N/A
Frequency Reuse Factor, 90%/99% Availability (Omni Cell)	7	7	7	7(4)	7	1 (total)
Omni Cell Range (m)	13,000 (max)	13,000 (max)	13,000 (max)	13,000 (max) (1600-8000)	10,000 (max)	13,000 (max)
Allocatable Bearer Capacity (kbps)	NA	19.843 per carrier	19.843 (3ch) per carrier	Speech/data 13/9.6	42 per carrier	9.6
Error Handling	NA	Yes	Yes	Yes (various)	Yes (4.5 kb/s)	Yes
Capacity (users/cell/1.25 MHz)	6	29.7-59	18	7.2	21	108 (3 sectors)

Table 8-5 Characteristics of Several Digital Cellular Standards

(a) Air-Interface Characteristics of Three Digital Cellular Standards

	Europe (ETSI)	North America (TIA)	Japan (MPT)
Access Method	TDMA	TDMA	TDMA
Carrier Spacing	200 kHz	30 kHz	25 kHz
Users per Carrier	8 (16)	3 (6)	3 (tbd)

(continued)

Table 8–5 Characteristics of Several Digital Cellular Standards (Continued)

(a) Air-Interface Characteristics of Three Digital Cellular Standards

	Europe (ETSI)	North America (TIA)	Japan (MPT)
Modulation	GMSK	π/4-DQPSK	π/4-DQPSK
Voice Codec	RPE 13 kb/s	VSELP 8 kb/s	tbd
Voice Frame	20 ms	20 ms	20ms[*]
Channel Code	Convolutional	Convolutional	Convolutional[*]
Coded Bit Rate	22.8 kb/s	13 kb/s	11.2 kb/s
TDMA Frame Duration	4.6 ms	20 ms	20 ms[*]
Interleaving	≈ 40 ms	27 MS	27 MS[*]
Associated Control Channel	Extra slot	In slot	In slot[*]
Handoff Method	MAHO	MAHO	MAHO[*]

[*] Ericsson proposal.

(b) Macro Cell Capacity of Cellular Standards

	Analog AMPS (ref)	GSM Full Rate	GSM Half Rate	ADC	JDC
Total Bandwidth, B_t	25 MHz	25 MHz	25 MHz	25 MHz	25 MHz
Bandwidth per Voice Channel, B_c	30 kHz	25 kHz	12.5 kHz	10 kHz	8.33 kHz
Number of Voice Channels, B_t/B_c	833	1000	2000	2500	3000
Reuse Factor, N	7	3	3	7	4
Voice Channels per Site, M	119	333	666	357	750
Erlangs per Square km (3-km site distance)	12	40	84	41	91
Capacity Gain	1.0 ref	3.4	7.1	3.5	7.6

(continued)

Table 8–5 Characteristics of Several Digital Cellular Standards (Continued)
(c) Technical Characteristics of Digital Cellular Standards

	GSM	ADC	JDC
Access Method	TDMA	TDMA	TDMA
Carrier Spacing	200 kHz	30 kHz	25 kHz
Users per Carrier	8 (16)	3	3
Voice Bit Rate	13 kbit/s (6.5 kbit/s)	8 kbit/s	8 kbit/s
Total Bit Rate	270 kbit/s	48 kbit/s	42 kbit/s
Diversity Methods	Interleaving Frequency Hopping	Interleaving	Interleaving Antenna Diversity
Bandwidth per Voice Channel	25 kHz (12.5 kHz)	10 kHz	8.3 kHz
Required C/I	9 dB	16 dB	13 dB

Table 8–6 Spectral Efficiency of Several Cellular Standards

STD	EFF	MOD./ACCESS
AMPS	0.33 b/s-Hz[*]	FM Mod. - FDMA
IS-54	1.62 b/s-Hz	π/4-DQPSK Mod. - TDMA
GSM	1.35 b/s-Hz	GMSK Mod. - TDMA
CDMA	0.983 chips/Hz (1.2288/1.25)	QPSK Mod. - CDMA

[*] Also referred to as modulation efficiency and bandwidth efficiency.

Table 8–7   Benefits from Using CDMA Cellular

- The same frequencies may be used in all cells, and signal orthogonality is manifested by pseudo-random coding techniques. The chip rate of the PN spreading sequence is about 1/10 of the total bandwidth allocated to each of the cellular service operators.
- Capacity is increased over analog systems by nominally a factor of ten.
- No frequency management or planning required since there is only one communication radio channel.
- More robust to interference and can operate with C/I in 10-dB range.
- Can combat multipath.
- No equalizer is required to correct for ISI caused by time delay spread such as used in TDMA systems (IS-54).
- Soft handover is possible since inter-cell movement does not require impulsive changes in frequency (channel change), like those required in IS-54 or AMPS. In CDMA cellular, mobiles communicate with two cells simultaneously.
- No time guard bands required as in TDMA, but requires RF guard bands.
- Voice activation where the activity factor of human voice is 35%. During listen time, interference is reduced and other channels may occupy this time.

- On the debit side: Power equality of signals from all mobiles in a cell must be maintained to prevent a single unit from usurping the cell site receiver. This has been referred to as the near-far problem.

8.6 Personal Communications Service

PCS is a small cordless phone that uses micro cells. It has many of the attributes of cellular phones to compete with macro cellular systems. The frequencies at which these systems will operate are indicated in Figure 8–10. The PCS service areas adopted are defined by the Rand-McNally commercial atlas, that is, 51 Major Trading Areas (MTAs) and 493 Basic Trading Areas (BTAs).

In November 1994, two standards organizations agreed to address the seven over-the-air standards that were proposed for PCS. These are listed in Table 8–8. Their parameters are shown in Table 8–9. It appears, however, that no single standard will be set. Notice that all systems use digital, and both TDMA and CDMA will be entertained. The TAG-5 standard was derived from the European DCS-1800 technology. A U.S. version known as DCS-1900 has been proposed as a U.S. standard. The first American PCS system operating in the Washington/Baltimore area (Sprint Spectrum) uses the European GMSK modulation similar to that used in their GSM cellular system (see Figure 8–11). It appears that the Europeans have gained a foothold in the U.S. via their well-entrenched GSM format used in Sprint Spectrum PCS. It is my opinion that the GSM-type operation will have an increasing presence in the U.S.

Roaming will only be possible if contiguous service areas use the same standard, or if not, dual-mode operation is utilized so that roaming will allow switching to a dif-

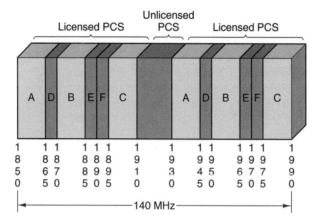

Channel Block	Frequency	Frequency	Total Bandwidth	Operators
MTA				
A	1850–1865	1930–1945	30	2
B	1870–1885	1950–1965	30	2
BTA				
C	1875–1910	1975–1990	30	4
D	1865–1870	1945–1950	10	4
E	1885–1890	1965–1970	10	4
F	1890–1895	1970–1975	10	4
Unlicensed				
	1910–1920 (data)		10	
	1920–1930 (voice)		10	

MTA: Major Trading Areas (51).
BTA: Basic Trading Areas (493).
MTAs & BTAs overlap, therefore up to six PCS operators per area.

• In Washington/Baltimore area, first PCS:
Sprint Spectrum: Block A, 1850–1865 (Mob to BS)
1930–1945 (BS to mob)

Figure 8–10 PCS channel plan.

ferent standard. It is unfortunate that PCS operators in the U.S. are considering different standards which will muddle inter-system roaming in the first decade of the twenty-first century.

In a *single*-cell system, CDMA is less efficient than FDMA or TDMA, since in an FDMA and TDMA cell, signals do not interfere with each other. In CDMA, all users in the cell *interfere with each other;* but in a cellular network, CDMA is more efficient.

Personal Communications Service

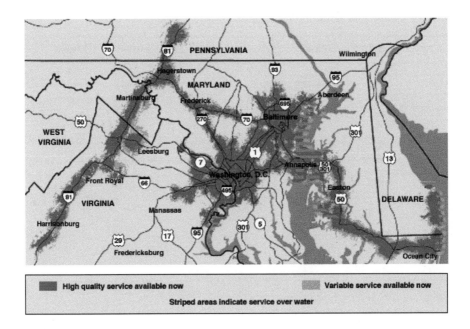

Figure 8-11 Sprint Spectrum PCS service areas. Copyright © 1988 Sprint Spectrum L.P. All rights reserved.

Table 8-8 Personal Communications System Standards

Personal Communication System (PCS) Standards (1900 MHz)	
TAG-1:	Composite TDMA/CDMA (Omnipoint)
TAG-2:	IS-95-based CDMA (QUALCOMM)
TAG-3:	PACS TDMA (Bellcore)
TAG-4:	IS-54-based TDMA
TAG-5:	DCS-based TDAM (GSM)*
TAG-6:	DCT-based TDAM (DECT)
TAG-7:	Wideband CDMA (InterDigital)

PACS: Personal Access Communications
DCS: Digital Cellular System
DCT: Digital Cordless Telephone
GSM: Global System for Mobile Communications

* Another version known as DCS 1900 has been proposed as a U.S. interim standard.

Table 8–9 Summary of Technical Characteristics*

	TAG-1	TAG-2	TAG-3	TAG-4	TAG-5	TAG-6	TAG-7
Parameter	New	IS-95-based	PACS	IS-54-based	DCS-based	DCT-based	W-CDMA
Access Method	CDMA/ TDMA/ FDMA	CDMA	TDM/ TDMA	TDM/ TDMA	TDMA	TDMA	D-CDMA
Duplex Method	TDD	FDD	FDD	FDD	FDD	TDD	FDD
Bandwidth	5 MHz	1.25 MHz	300 kHz	30 kHz	200 kHz	1728 kHz	5 MHz
Bit Rate (no overhead)	32 kb/s	8/13.3 kb/s	32 kb/s	7 kb/s	13 kb/s	32 kb/s	32 kb/s
Process Gain	21 dB	21 dB	NA	NA	NA	NA	21 dB
Channel Spacing	5 MHz	1.25 MHz	300 kHz	30 kHz	200 kHz	1728 kHz	5 MHz
Voice Channels/ Carrier32 (8 kb/s CELP) SHO = soft handover	20 (eff) + SHO	8	3	8	12	128 (less SHO)	
Reference to AMPS	16 X	10 X	0.8 X	3 X	2-3 X	0.2 X	16 X (less SHO)
Modulation	Cont. ph. shift QM	OQPSK/ QPSK	$\pi/4$-DQPSK	GMSK	GFSK	OQPSK/ QPSK	
Error Control (voice)	None	FEC	None	FEC	FEC	None	FEC
Frequency Reuse (N)	3	1	16 × 1	7 × 3	7 × 1 and 3 × 3	9	1
Max. Avg. Subscriber Power	–	200 mW	12.5 mW	100 mW	125 mW	20.8	500 mW
SU Power in Timeslot	1 W	–	100 mW	600 mW	1 W	250 mW	–
Time Frame Length	625 ms	–	312.5 ms	6.7	577 ms	417 ms	–

Table 8–9 Summary of Technical Characteristics*

	TAG-1	TAG-2	TAG-3	TAG-4	TAG-5	TAG-6	TAG-7
Timeslot Length	80 ms	50 ms	9 ms	110 ms	90 ms	28 ms	13.25 ms
End-to-End Speech Delay	80 ms	50 ms	9 ms	110 ms	90 ms	28 ms	13.25 ms
Equalizer	No	No	No	Yes	Yes	No	No
Vocoder	CELP (8 kb/s) ADPCM (16, 24, 32, 40 kb/s)	Var. rate (8/4/2/1)	ADPCM (32 kb/s)	VSELP (8 kb/s) LDCELP (16 kb/s)	RPE-LTP (13 kb/s)	ADPCM (16-32 kb/s)	ADPCM (32 kb/s)

* Taken from a presentation given at the National Engineering Consortium WPC Teleforum III. C. I. Cook, "Development of Air Interface Standards for PCS," *IEEE Personal Communications*, 4th Quarter 1994. Copyright © 1994 by the Institute of Electrical and Electronics Engineers, Inc. Reprinted with permission.

8.7 Conclusions

America is starting to become unwired. The insatiable desire to go wireless has caused a shortfall in capacity reflected by the entrenched analog cellular system, AMPS. As the number of subscribers proliferates, methods must be devised to satisfy this demand by improving existing systems and/or introducing new technologies.

In pursuit of this goal, increasing cell density has been achieved without an increase of additional spectrum. However, cell splitting is expensive in terms of base stations and the real estate required for the base stations. In addition, the sprouting of base station antennas all over the place has not been looked at favorably by the general public.

A more benign and viable approach is being exploited by going from analog modulation to digital modulation. Digital has several virtues, but improvement in voice quality is not necessarily one of them. The real reason for going to digital is to increase capacity, not to improve voice quality. In fact, if source compression in the codec goes too far (to get more capacity), voice quality may suffer significantly. It is my understanding that IS-54 is experiencing degraded voice quality at 8 kbps. Nevertheless, digital modulation using TDMA can increase capacity by at least a factor of three. It has been further demonstrated that the CDMA system, IS-95, can realize capacity improvement of greater than a factor of ten.

Some problems that are continuing to plague wireless designers are the multipath and shadowing problems, and their deleterious effects on performance. As is well-known, generally there is no line-of-sight propagation between the base station and mobile, and normally log-normal-Rayleigh fading is experienced. The use of digital modulation can reduce part of this problem by incorporating FEC and interleaving. In CDMA systems, additional improvement can be realized by RAKE receivers.

Other methods are being used to enhance capacity and are indicated in Table 8–1. All provide varying degrees of performance, complexity, and cost. Probably the greatest breakthrough in wireless communications was the cellular concept.

It is unfortunate that both cellular and PSC standards in the U.S. have taken different paths and therefore are incompatible. Conversely, the European community of 16 nations (CEPT) has settled on one standard, the GSM. This therefore permits roaming throughout Europe.

Even with the multiple standards in the U.S., there is some solace in the fact that multimode mobile units are being made which can accommodate AMPS/TDMA and AMPS/CDMA. It is unlikely that tri-mode sets will come to fruition, but only time will tell.

Because of the paucity of useable spectrum, innovative technology is being developed in order to increase the capacity of wireless systems. One avenue which is being pursued is smart antennas. That is, antennas which take advantage of an antenna beam directionality in order to reduce interference and increase frequency reuse, even within a cell.

Chapters 9–14 will deal with emerging smart antenna technology. One important network which is playing an increasing role in one version of smart antennas is the Butler matrix. This is basically a labyrinth of fixed RF phasers and hybrid junctions to form beams in space. The Butler matrix is described in more detail in Chapter 9.

8.8 References

[1] FCC, "Notice of Proposed Rule Making for Emerging Technologies at 1850-2000 MHz," ET Docket 92-9, January 16, 1993.

[2] W. R. Young, "AMPS-Introduction, Background and Objectives," BSTJ, January 1979.

[3] V. H. Mac Donald, "AMPS-The Cellular Concept," BSTJ, January 1979.

[4] B. Pattan, "Spectral Efficiencies Through the Expedient of Modern Digital Communications," FCC/OET Internal Report, February 1995.

[5] R. Padovanni, "Reverse Link Performance of IS-95 Based Cellular System," *IEEE Personal Communications*, 3rd Quarter 1994.

[6] A. J. Viterbi, "The Orthogonal-Random Waveform Dichotomy for Digital Mobile Personal Communication," *IEEE Personal Communications*, Vol.1, No. 1, 1st Quarter 1994.

[7] I. Kalet, "Digital Cellular and PCS Communications," School notes of course taken at George Washington University, 1995.

[8] W. C. Y. Lee, *Mobile Communications Engineering*, McGraw-Hill Book Co., New York, 1982.

[9] G. Calhoun, *Digital Cellular Radio*, Atech House, Norwood, MA, 1988.

[10] "Wireless Personal Communications," *IEEE Communications Magazine*, January 1995.

CHAPTER 9

The Butler Matrix

9.1 Introduction

Both the IRIDIUM (LEO) and ODYSSEY[1] (MEO) mobile satellite systems use constellations to provide worldwide continuous coverage. Each satellite in the constellation produces antenna beams covering wide areas. However, because the coverage areas are large, and the users in a coverage area may exceed the capacity of a single beam, the coverage is blanketed by multiple spot beams, which approximate a cellular structure on the ground. This, like in terrestrial cellular systems, permits frequency reuse, thus enhancing the capacity of the communication systems. Clearly, additional means can be used to enhance capacity, but this chapter will confine its attention to only cell topology. The benefits accrued from using cellular partitioning are achieved via spot beams. These benefits also result from the fact that the ground transceiver sees a higher EIRP emanating from the satellite, as well as a higher G/T in transmission mode.

Spot beams are generated by the on-board multi-beam antenna sub-system, which generates and lays down multiple spot beams, approximating a cellular structure on the surface of the earth. The multi-beam antenna design used by both IRIDIUM and ODYSSEY uses a two-dimensional *Butler matrix* beam-forming network to synthesize a pincushion footprint over the desired coverage area.

The Butler matrix has been used extensively over the years in radar, electronic warfare (electronic support measures), and satellite systems [3]. The Butler matrix consists of passive four-port hybrid power dividers and fixed phase shifters. It has N input

---

1. In 1998, ODYSSEY decided to merge with Intermediate Orbit Communications (IOC), which also is an MEO satellite system.

ports and N output ports. As a beam-forming network, it is used to drive an array of N antenna elements. It can produce N orthogonally-spaced beams, overlapping at the −3.9 dB level and having the full gain of the array. The network is most commonly used to produce beams in one plane, but can be designed to produce volumetric beams in a pincushion deployment.[2]

9.2 Planar Array Beams

A generic version of the Butler matrix used as a beam-forming network is shown in Figure 9–1. It consists of a 3-dB quadrature hybrid [1] (see Figure 9–2) driving two antenna elements with separation d. Note the amplitude and phase relationships for the hybrid structure in Figure 9–2. For example, in Figure 9–1, feeding the lower left-hand port of the coupler results in the antenna array being uniformly illuminated and differentially phased, which points the resulting beam peak to the right of boresight in direction "Q". With the phasing indicated, we obtain

$$(2\pi d/\lambda)\sin\theta_1 - \pi/2 = 0$$

$$\text{or} \quad \sin\theta = \lambda/4d$$

where d is element separation. Feeding the lower right port, 2, results in a beam pointing to the left of boresight at an angle $\sin\theta_2 = -\lambda/4d$. If both ports 1 and 2 are driven, two beams will be produced at angles $\sin\theta_2 = \pm\lambda/4d$. The matrix produces N (equals 2 in our example) orthogonally-spaced beams overlapping at the −3.9 dB level.

A slightly more detailed version producing four beams is shown in Figure 9–3. This consists of $N = 4$ quadrature hybrids and fixed phase shifters. Notice that if we trace a phasor through the network, no boresight beam is formed and the beams are symmetrically deployed about the array axis. For the ports driven as indicated in the figures, we see that the phase front across the aperture elements is −45°, −90°, −135°, and −180°. Therefore, the shaded beam is produced.

The matrix depicted in Figure 9–3 may also have a sequencing switch on the input ports to scan any one of the four positions shown. Clearly, a more elaborate labyrinth can scan and produce many more beams.

In the basic array, the number of beams is equal to the number of elements, and the array factor is of the form $\sin Nx/\sin x$ where N is the number of elements. Kraus [10] has shown that the magnitude of the field intensity in the far field of a linear array of N isotropic radiators is given by

---

2. To be orthogonal, SinX/X patterns must be spaced so that the crossover is at about 4 dB down ($2/\pi$), and the sidelobes are down 13.2 dB. Tapering will violate this requirement. For example, for cosine taper, the crossover level becomes about −9.5 dB for orthogonality.

Planar Array Beams

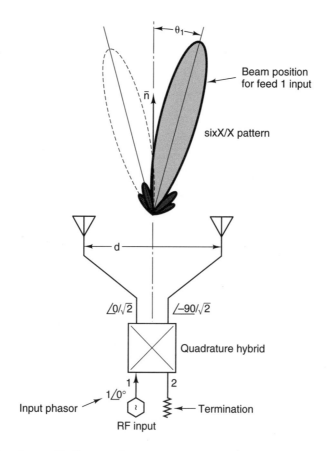

Figure 9–1 Two-beam Butler matrix.

$$E(\theta) = E_e(\theta) \frac{1}{N} \frac{\sin(N\phi/2)}{\sin(\phi/2)}$$

where $E(\theta)$ is the element factor (weighs the array factor), ϕ is $(2\pi d/\lambda)\sin\theta - \delta$ radians, δ is the progressive phase difference generated by the matrix and is equal to $\delta k = (2k-1)\pi/N$, $k = 1, 2, ..., N/2$, and k is the beam number.

Note in the two-element array shown in Figure 9–1 that

$$\delta_k = (2k-1)\pi/N$$
$$\delta = \pi/N = \pi/2$$

The location of the beams can be found from the following relationship:

$$\sin\theta = (\pi/Nd)[k - (1/2)]$$

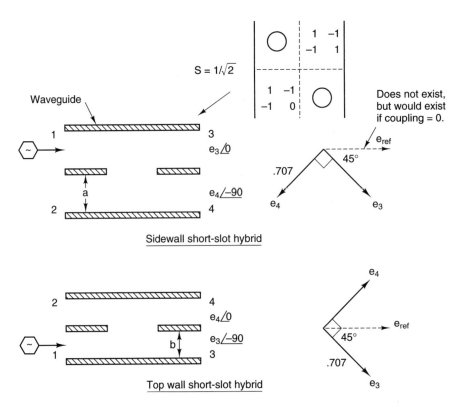

Figure 9–2 Phase relationships in riblet short-slot hybrids, the basic building blocks of the Butler matrix utilizing a waveguide implementation.

The first sidelobe is down 13.5 dB, which is typical for a linear array with equal amplitudes and equal spacing. Interestingly, the beams cross over at the 3.9-dB points, which suggests that the beams are orthogonal and the network is lossless.

The sidelobes can be significantly improved by coherently combining two output ports to give a cosine variation with the sidelobes down by 23 dB. However, a beam distention of 35 percent results. Figure 9–4 illustrates the addition of two beams, each having uniform illumination (with $\sin x/x$ far field) to obtain a cosine illumination.

Other tricks can be manifested by the matrix. The beams can be made to have limited scan by the use of the network shown in Figure 9–5. Power is varied to ports 3 and 4 (say), and the variable power divider controls the power to each port. Varying the division of power steers the beam between individual beam axes.

Multiple Volumetric Beams

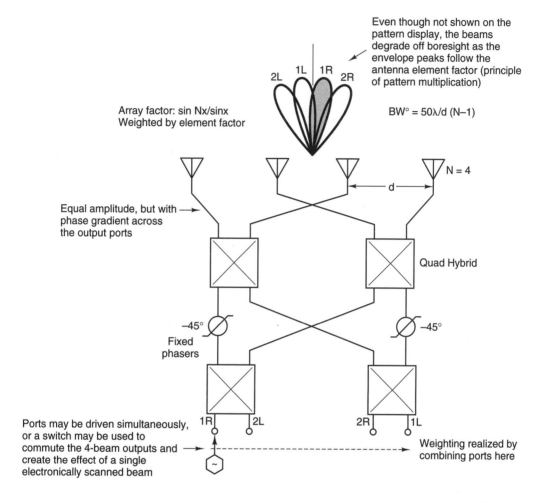

Figure 9–3 Four-beam Butler matrix.

9.3 Multiple Volumetric Beams

Multiple volumetric beams can be generated by a Butler matrix by dividing the array into rows and columns. This is depicted in Figure 9–6. The columns consist of

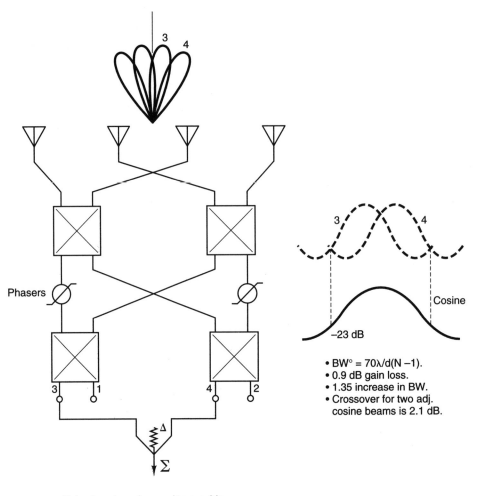

Figure 9–4 Cosine illumination achieved by combining two adjacent beams. In general, $n + 1$ beams can be added to form a $(Cosine)^n$ illumination.

linear Butler matrices, where the outputs of each column are the vertical number of beams. For this example, four beams are in each vertical stack. The outputs of the column matrices drive four row matrices. These are also Butler matrices. The outputs of the row matrix are used to produce four squinted vertical stacks. Note that prior to squinting, the *column* matrices produce vertical beam stacks (four per stack), which all point in the same direction.

Multiple Volumetric Beams

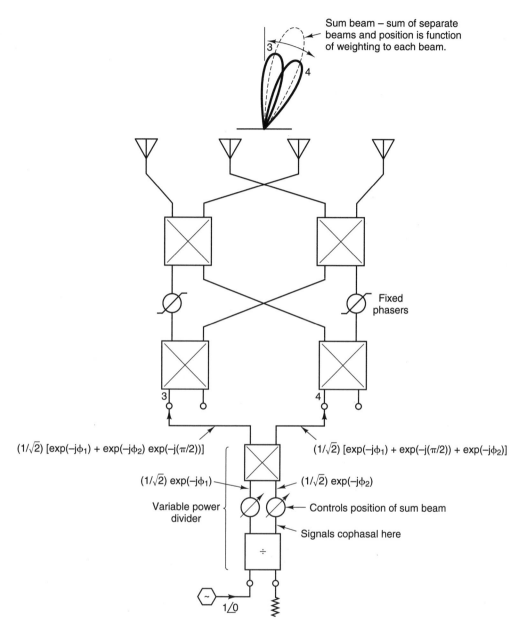

Figure 9–5 Beam scanning in a Butler matrix using a variable power divider.

For example, beam 1 is formed by column matrix 1, but is squinted upward from the horizontal boresight plane by the inherent operations of the matrix. The beam is

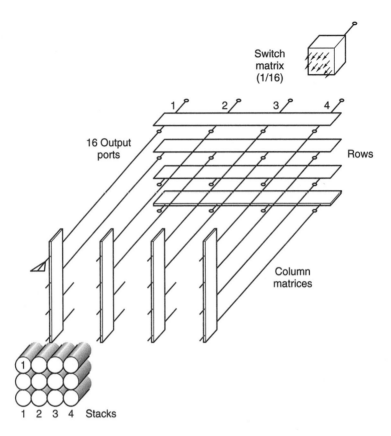

Figure 9–6 Two-dimensional Butler matrix producing orthogonal beams.

now squinted to the left from the vertical boresight plane by the phasing of row matrix 1 and its port, 2. Correlating Figure 9–3 with Figure 9–6 will help to make this clear.

There is a practical problem associated with the system shown in Figure 9–6. The stacked beam will form pencil beams only if the aperture size, in the direction perpendicular to the column matrices, is comparable to the array lengths. If there are horns with a cosine distribution, the aperture width (w) in the direction would have to be equal to $w = 69\lambda / BW$, where BW is the beamwidth in the vertical direction. The beamwidth in the vertical direction results from the array length. In the orthogonal direction, a very large flare may be required.

The generation of beams for volumetric coverage by a Butler matrix labyrinth requires considerably complex circuits. Additional complexity results from circuitry to reduce the antenna beam sidelobe levels with attendant low crossover (9.5 dB for cosine taper or lower still for a smaller sidelobe level) or high loss with higher crossover.

Multiple Volumetric Beams

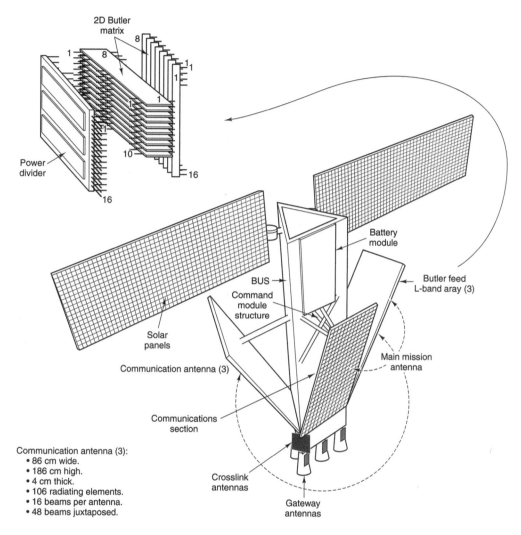

Figure 9-7 Sketch of the IRIDIUM satellite design.

The Butler matrix requires

$$H = (N/2)\log_2 N = (N/2)(\log_{10} N / \log_{10} 2) = 3.32(N/2)\log_{10} N$$

quadrature couplers, plus fixed phase shifters. The number of fixed phase shifters is

$$P = (N/2)\log_2 N - 1$$

For example, for a 64-beam system (8×8), 16 planar matrix feeds are required with 12 hybrid junctions and 16 fixed phasers for each matrix.

9.4 Butler Array Application

A sketch of the IRIDIUM spacecraft using the Butler matrix analog beam-forming network is shown in Figure 9–7. Each of the three communications signal arrays generate 16 spot beams. These beams are juxtaposed to produce 48 beams in the coverage area.

9.5 Conclusions

The Butler matrix is a versatile device. It can serve as a beam-forming network, permitting volumetric beams to be generated which are orthogonal and independent, and each port will have the gain of the full array. Being both passive and reciprocal, they can be used for reception *and* transmission in an antenna array. The beams may be deployed simultaneously or sequentially, depending on the application.

9.6 References

[1] H. J. Riblet, "The Short-Slot Hybrid Junction," *Proc. IRE*, February 1952 (there is a phasing error in Riblet's paper)

[2] J. Butler and R. Lowe, "Beam-Forming Matrix Simplifies Design of Electrically Scanned Antennas," *Electronic Design*, April 12, 1961.

[3] J. P. Shelton and K. S. Kelleher, "Multiple Beams From Linear Arrays," *IRE Trans. Ant. and Propagation*, March 1961 (may also have independently invented the matrix).

[4] T. Macnamara, "Simplified Design Procedure for Butler Matrices Incorporating 90 and 180 Hybrids," *IEE Proc.H, Microwave and Antenna and Propagation*, pp. 50-54, 1987 (British).

[5] J. R. F. Guy, "Proposal to Use Reflected Signals Through a Single Butler Matrix to Produce Multiple Beams From a Circular Array Antenna," *Electronic Letters*, 1985, pp. 209-211.

[6] R. Levy, "A High Power X-Band Butler Matrix," *Microwave Journal*, April 1984.

[7] J. P. Shelton, "Multibeam Planar Arrays," *Proc. IEEE*, November 1968 (aside: B. Pattan also has an 11 page article on Phased Arrays in this issue).

[8] R. C. Hansen, *Microwave Scanning Antennas, Volume III*, Academic Press, New York, 1966.

[9] *Microwave Development Laboratory Inc. Handbook*, Matick, Mass.

CHAPTER 10

Sidelobe Cancellers in Smart Antenna Applications

10.1 Introduction

One of the earliest forms of quasi-adaptive generic arrays was the sidelobe canceller (SLC). It had its genesis in the radar field and was used by defense radars to combat jamming into the radar antenna sidelobes. In a more benign environment, it is used in GSO satellite applications to mitigate interference to earth stations from terrestrial systems such as fixed service transmitters. More recently, it has been considered as a possible remedy to reduce interference to cellular systems, in particular, those using sectorization and smart antennas.

This chapter will briefly attempt to explain the workings of sidelobe cancellers. These may take on several configurations operating at RF, IF, or even video frequencies. We will confine our discussion to those operating at IF frequencies, and these are of the Howells' genre [1].

10.2 Single Interferer Sidelobe Canceller

As the name suggests, a sidelobe canceller attempts to suppress the interference coming into the sidelobes of a directional beam antenna used in the communication channel. A typical geometry of a sidelobe canceller is depicted in Figure 10–1, in which one interferer is cancelled. The sidelobe canceller can exist in several forms, but as alluded to above, we will focus on an IF sidelobe canceller.

Two channels are required in the system. The signal channel uses a directional receive antenna, which picks up a signal from the mobile cell. As a realizable antenna,

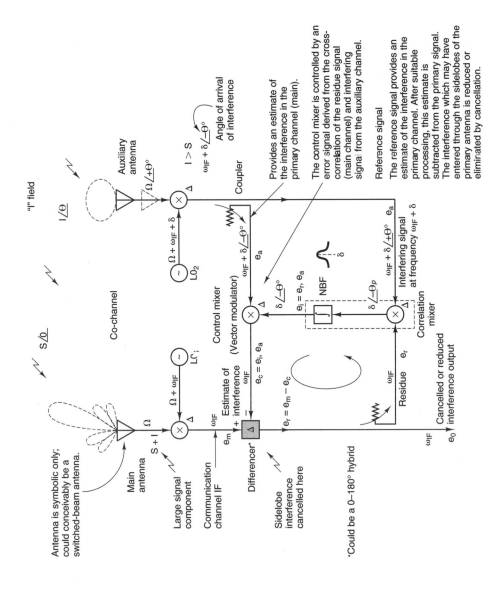

Figure 10–1 Mechanics of a single-loop IF sidelobe canceller.

it will have sidelobes which are vulnerable to interference illuminating the sidelobes. Clearly, this degrades the carrier-to-noise ratio.

The other channel is the auxilary channel, and its role is to pick up the interference signal and discriminate as much as possible against the bona fide signal.[1] The interference must exceed the desired signal in this channel. Its antenna beam is generally broader and wider than the signal beam, and its gain is larger than the gain of the signal antenna sidelobes.

Similarly, for the signal channel, the signal will be larger than the interference, in part due to the directionality of the antenna in the direction of its desired signal. Clearly there is some interference in the signal channel coming through its sidelobes. This is what we are attempting to suppress.

Both the incoming signal and interfering signal incident on their parent antennas will be heterodyned down to a frequency at IF. In the signal channel (see Figure 10–1), the input frequency, Ω, is heterodyned down to IF, ω_{IF}, using a local oscillator frequency, $\Omega + \omega_{IF}$. In the auxilary channel, the interfering signal, which is coming in at angle θ, is heterodyned down to $\omega_{IF} + \delta + \angle-\theta°$, where δ is a frequency which is somewhat greater than the bandwidth of the signal.

In the signal channel, a differencer network compares the interference coming down the channel with the interference from the control mixer and cancellation takes place. However, this cancellation is not complete, and some interference residue exists. The differencer is sampled and is used as an input to the correlator.

In the reference channel, the interference is sampled and serves as the second input to the control mixer. It further serves as the second input to the correlation mixer. The interference in the residue correlates with the interference from the reference channel.

The phase and amplitude information needed to control the control mixer is obtained by performing a correlation between the interfering residue (i.e., the interference remaining after cancellation) and the auxilary channel interference, which provides an interference phase difference. The correlation mixer detects the presence of the interference in the receive channel and adjusts the signal controller adaptively to minimize the interference. There is a bona fide signal which enters the correlator from the main channel (via the coupler), but this is uncorrelated with the interference coming from the auxilary channel, and does not contribute a deterministic signal to its output.

At the output of the correlation mixer, the signal is integrated (narrowbanded)[2] and centered on the difference between the IFs of the two channels. This signal serves as a local oscillator signal for the control mixer. At the control mixer, the signals are heterodyned to the signal channel's IF with the attendant interference. These signals

---

1. Actually, the auxilary channel is to provide an independent replica of an interference signal in the sidelobes of the main channel beam pattern for cancellation.
2. A narrowband filter is an integrator.

drive the differencer where cancellation or subtraction of the interference in the signal channel and looped supplied interference takes place. Notice that the phase difference between the signal channel and direction angle of the interference is cancelled by the control mixer.

In the correlation mixer, the difference frequency is utilized and there is a conjugation of the phase. The principle of phase conjugacy is invoked in the mixing process. The lower sideband (designated by the symbol Δ) delays the phase.

To summarize, the subtraction network, correlation mixer, narrowband filter, and controller mixer form a nonlinear closed loop that provides automatic gain adjustment so that the residual interference in the signal channel approaches zero.

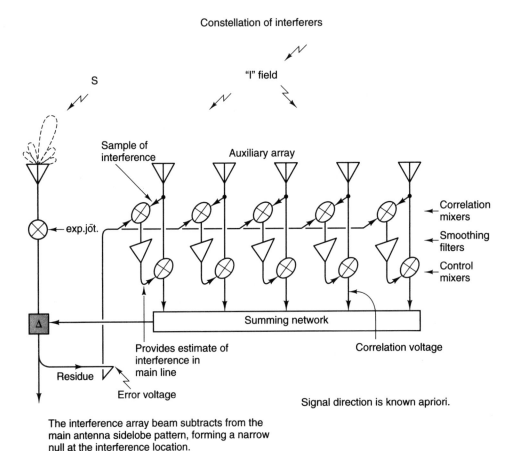

Figure 10–2 Multiple (5) adaptive sidelobe cancellations in an interference field.

10.3 Multiple Interferers

The SLC described above is used to suppress a single interferer. The single canceller's performance against multiple interferers is limited by the fact that it has only one degree of freedom to use against N-dimensional interferers. A sidelobe canceller can be designed to perform against several interferers. This is achieved by using additional auxilary antennas, with each antenna serving to reject an interferer. A simplication of a multi-interference canceller is shown in Figure 10-2 [2]. Note that each auxilary channel has its own feedback loop consisting of correlation mixer, integrator (smoothing filter), and control mixer. The unit shown can cope with five interferers coming from the general direction of the bona fide signal.

10.4 Conclusions

Since sidelobe cancellers appear to have application in smart antennas to remove interference entering the sidelobes of the communication antenna, they can be used to complement switched-beam antenna systems to combat interferers, even within the cell. Depending on the complexity (and cost) permitted, cancellers can reject several interferers.

However, cancellers are not able to reject an interferer in the communication channel beam. The interferer must always be bigger than the bona fide signal to realize a useable signal out of the correlator, which serves as an error signal. The antenna gain of the auxilary antenna is greater than the sidelobes of the communication channel.

Notice that sidelobe cancellers are not truly adaptive since they utilize a fixed beam. An antenna null is effectively placed in the direction of the interferer(s).

10.5 References

[1] P. W. Howells, "Intermediate Frequency Side-Lobe Canceller," Patent No. 3,202,990, August 24, 1965.

[2] Special Issue on Adaptive Antennas, *IEEE Trans. on Antenna & Propagation*, September 1976.

[3] S. P. Applebaum, "Adaptive Arrays," Syracuse Univ. Research Corp. (SURC), Rpt 66-001, August 1966.

[4] T. A. Bristow, "Application of an Interference Cancellation Technique to Communication and Radar Systems," *Systems Technology*, September 1979.

[5] R. Nitzberg, "Canceller Performance Degradation Due to Estimation Noise," *IEEE Trans. on Aerospace & Electronics Systems*, September 1981.

[6] S. C. Swales et al., "The Performance Enhancement of Multibeam Adaptive Base-Station Antennas for Cellular Land Mobile Radio Systems," *IEEE Trans. Vehicular Technology*, February 1990.

CHAPTER 11

A Look at Switched-Beam Smart Antennas

11.1 Introduction

Increasing the capacity of a cellular system, in the presence of limited spectrum, is a never-ending pursuit. Methods have been devised which provide capacity enhancement. These include:

- Cell splitting.
- Converting to digital with source compression (with small reduction in voice quality).
- Cell sectorization.

As the name suggests, cell splitting refers to dividing up a hexagonal cell into a multiplicity of smaller cells, each with its own base station. Cell splitting increases the number of channels per unit area. This is feasible and is being done, especially in urban areas. This also brings along a host of problems, in particular, the greater number of hand-offs required, but probably more important is the additional cost of the base stations and the acquisition of real estate on which to locate the cell sites.

Sectorization, in which a cell is divided, for example, into 120° sectors, is also being used. This implies that each sector is served by a directional antenna with a nominal beamwidth in azimuth of 120°. Actually, an ideal sector beam, with no spillover or sidelobes into other sectors, is not physically realizable.

From an interference point of view, in a common seven-cell cluster that uses omni-directional antennas, six cells in the sea of cells provide interference to Cell 1. This is made clearer in Figure 11–1. On the other hand, the 120° sector beam uses only

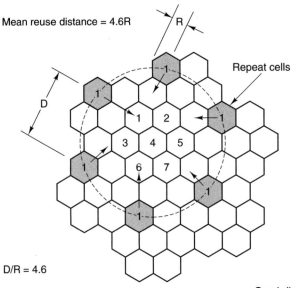

- D/R: co-channel reuse ratio.
- D: separation between two co-channel cells.
- R: radius of cells.
- In hex-shaped cellular mosaic, D/R = √3N.
- N: number of cells in cluster

Figure 11–1 Ideal seven-cell cluster repeat pattern.

two of these peripheral cells to provide interference. This is shown in Figure 11–2. It is clear that the four other cells will have much lower sidelobes or backlobes illuminating Cell 4 in the cluster. We can also extend this sectorization concept to 60° sectors, as shown in Figure 11–3. Here, only one peripheral cell sends main beam energy in the direction of the victim cell.

For the case where omni-directional coverage is used, the required *C/I* bandied around the industry, to achieve an adequate voice quality, is about 17–18 dB. The bona fide signal is equal to the signal received by the mobile in the affected cell, divided by the sum of the interference from the offending cell site's operating co-channel. That is, the signal-to-interference (*S/I*) ratio is given simply by

$$S/I = S/N_{thermal} + \sum I_i^*$$

If we assume the propagation loss due to range, *R*, is the same for the interferers and the signal and power radiated is the same, the *S/I* can be approximated by

Introduction

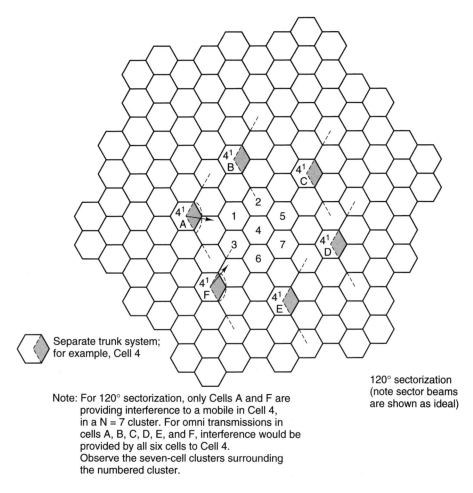

Figure 11–2 Cellular structuring showing 120° sectorization.

$$S/I = R^{-n} / \sum (D_i)^{-n}$$

where n is typically between 3 and 4 (no line-of-sight propagation), R ranges from base station to its parent mobile, and D is the distance from interferer(s) to mobile.

Using sectorization, there will be an average of 6-dB improvement in S/I, resulting from exercising the expression above. However, on the debit side, there is a decrease in capacity. However, this decrease in capacity can be made up by decreasing the cluster size, or the D/R ratio as defined in Figure 11–1 for the omni case. For a four-

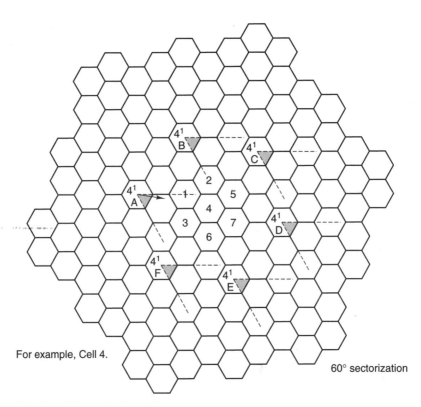

For example, Cell 4.

60° sectorization

Note: For 60° sectorization, only one interferer radiates into Cell 4 of the N = 7 cluster.

Figure 11-3 Cellular structuring manifesting 60° sectorization.

cell cluster with 120° sectorization, the D/R is 3.46. For a 120° sectorized cell, the four-cell cluster is shown in Figure 11-4.

Sectorization reduces trunking efficiency since the number of users per sector that can access the system is decreased. The pool from which the number of random users can select is restricted. It has been shown that the trunking efficiency is reduced about 25% from the unsectored cell. However, in toto, this efficiency can be improved by reducing the cluster size from seven to four.

An additional attribute of sectorization is the increase in gain from the directional beams. It allows the power to be reduced on the downlink while increasing the signal reception on the uplink. On the debit side, the increase in segmentation of the cells will

Introduction

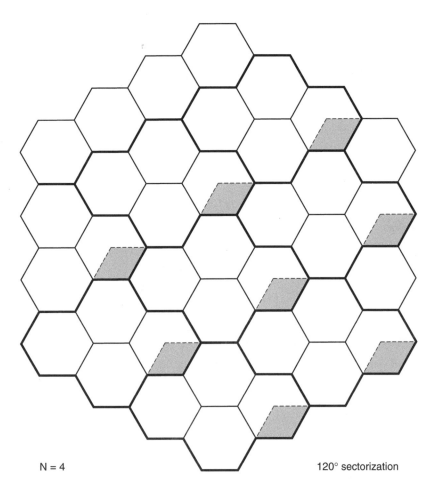

N = 4   120° sectorization

A cell cluster is outlined in bold and
replicated over the coverage area.

Figure 11-4 A 120° sectored cell which is straddled by a 120° beam width directional antenna beam.

require a greater number of hand-offs as the mobile moves from one sector to another within the cell, in addition, of course, from cell to cell.

Extending the sectorization one step further to a six-sector cell, there is a further improvement in the *D/R* with a further decrease in trunking efficiency. There is now one interferer illuminating the center cell in the seven-cell cluster, which offers a *C/I* of about 29 dB, which is substantially above the omni coverage case.

The benefit accrued in the S/I from 120° sectorization depends on some rather ideal conditions, such as:

- Each interfering site is always transmitting.
- The antenna patterns are ideal sector beams with no spillover into other sectors, and no side- or backlobes.
- Each site is at the same height.
- Each site has a 40 dB per decade ($1/R^4$) propagation loss.
- Each interfering site is equal-distanced from the victim site.
- Only two sites interfere.

11.2 Trunking Efficiency

In terrestrial cellular systems, both the downlink (base station-to-mobile) and uplink (mobile-to-base station) use 25 MHz of bandwidth. For analog AMPS, the channel spacing is 30 kHz. Therefore, there are 25MHz / 30kHz = 840 channel pairs which are divided between two operators.

In a cluster of seven cells, each cell uses one-seventh of the channels (neighboring cells do not use the same frequencies). Therefore, each cell uses 840 / 7 = 120 channel pairs per cell per 25 MHz, or 60 channels per operator. Some of these are control channels and do not supply service; for simplicity we will neglect this fact here, even though it normally would be considered in the calculations.

If 120° sectorization is used, each operator will have access to 60 / 3 = 20 channels per sector [3]. For 1 percent blocking, one can use the Erlang B traffic tables to determine the Erlang traffic per sector (see Appendix C). These tables have also been referred to as the Blocked-Calls-Cleared Tables. From the table given in Appendix C, for 20 channels and 1% probability of blocking, the traffic loading is about 12 Erlangs per sector, or a total of 36 Erlangs. For the omni case and 60 channels per cell, the traffic loading is 46.9 Erlangs.

We therefore witness a reduction in trunking efficiency in going from omni coverage to sectorization. However, as indicated previously, there is an improvement in C/I of about 6 dB, which allows flexibility in reducing the reuse factor (decreasing the D/R). Smaller cell clustering will increase capacity.

11.3 Smart Antennas

Smart antennas are basically an extension of beam sectorization in which the sector coverage is supplanted by multiple beams. This is achieved by the use of array structures, and the number of beams in the sector (e.g., 120°) is a function of the array extent.

The increase in beam directionality can provide an increase in capacity and expand cell site geographic coverage. The former is more applicable to urbanized areas and the latter to rural areas. At the mobile end (uplink mode), the benefit accrued by reducing transmitter power, because of the greater gain of the base station antenna, can extend the life of the battery.

The increased rejection of interference by the use of multiple beams: (1) permits a tighter reuse structure; (2) can improve the quality of voice communications through the vehicle by an increase in C/I (>18 dB); and (3) can reduce multipath, thus tempering the power margin requirements.

There are basically two types of smart antennas which are being investigated by independents or cellular companies to enhance the performance of their systems. Probably the better word to use in lieu of performance is increased capacity. The antennas are of switched-beam and adaptive array types. The first consists of generating a multiplicity of juxtaposed beams whose output may be switched to a receiver or a bank of receivers. The cell is therefore blanketed with a cluster of contiguous beams. The second approach to smart antennas is via the use of adaptive arrays. In adaptive arrays, the beam(s) structure adapts to the RF signal environment and directs beams to bona fide signals, depressing the antenna pattern(s) in the direction of the interferers. Adaptive arrays are generally more digital-processing-intensive than switched-beam systems, and are more expensive. We might also add that the theory to explain adaptive arrays is more mathematically intensive. An early version of an adaptive array is a sidelobe canceller, which was described in the preceding chapter.[1] It was first used in the military sector, but now has commercial applications. It was also used in communications satellite systems, at the earth station antenna, to remove terrestrial interference.

In both versions, antenna array configurations are used and all beams generally utilize the full array for beam formation. Signals from the array elements are added coherently at the output.

Switched-beam systems deploy a fixed set of relatively narrow azimuthal beams, which are generally narrower than those used in 120° sectorized systems. The RF output to the beams is either RF or baseband digitally processed to ascertain the sector in which the communicating mobile may be located. The cellular space is broken down into three sectors (120°), with each sector served by an array of radiating elements fed by a beam-forming network, which ideally forms independent beams. A typical display of these beams at a cell site is shown in Figure 11–5. The figure shows six beams in each 120° sector, with a nominal beamwidth of 20° in azimuth.

---

1. Actually, the SLC is not fully adaptive (it has a fixed beam).

• A switched beam or fixed beam smart antenna system.

Figure 11–5 A rosette of directional beams for a smart antenna.

Another topology of beams is shown in Figure 11–6. Here there are four beams per 120° sector with 12 beams for 360° coverage. Each beam is about 30° wide.

For both of these beam deployments, each sector is served by an array panel in a triangular arrangement (could be a conformal array). The array size for the twelve-beam system is about 2.5 feet. The current cellular towers use triangular platforms to

Smart Antennas

Beamwidth: ≈ 30° nominal
Aperture size: ≈ 0.8 meters per sector (at 900 MHz).

Figure 11–6 A twelve-beam system for omni coverage.

support the antenna farm. These platforms are four meters on each side, and therefore would have no problem in accommodating the array structures.

For a linear array, the beamwidth in azimuth is determined by the array length and wavelength. In the vertical direction, the antenna elements (at UHF these may be log periodics) are stacked to reduce the beamwidth in this direction. This will also increase the antenna gain, since the antenna gain is a function of both the elevation and azimuth beamwidths. To the first approximation, the gain may be given by $G = 40,000\eta / \theta,\phi$, where η = antenna efficiency and θ and ϕ in degrees are beamwidths in azimuth and elevation, respectively.

The most popular network to form the beams in switched-beam technology is the Butler matrix. This network was briefly described in a previous chapter, but we will review it again here and add some subtle points regarding its operation.

The topology of the Butler beam-forming matrix for an eight contiguous beam system is shown in Figure 11–7. There are eight input ports and eight output ports ($N = 8$). This is a reciprocal structure, thus either end can be the RF input or RF output. The matrix consists of quad hybrids, or directional couplers and interspersed passive fixed phase shifters. The amount of each depends on the number of beams generated. For example, for a linear array of N elements, the number of couplers is

$$N = (N/2) \log_2 N \tag{11.1}$$

where N is the number of beams., and the number of fixed phase shifters is

$$N = (N/2)(\log_2 N - 1) \tag{11.2}$$

For a large number of ports (or beams), the number could be quite large. However, in cellular applications, the number will be modest.

Butler arrays can be built to have any power of two beams: 2, 4, 8, 16, 32, and so forth. The number of beams is equal to the number of array elements (N). Beam-forming techniques can be used in two-dimensional (planar) arrays by first combining the output of the columns of antenna elements into matrices and then combining the output of the column matrices into a group of row matrices.

At midband, where the antenna element spacing is one-half wavelength, the beam positions are given by

$$\mathrm{Sin}\theta = 2k - 1 / N$$

where θ is the angle off boresight and k is the beam number.

The beamwidth and beam spacing vary inversely with frequency, thus maintaining the same crossover (at –3.9 dB).

For a practical array operating at about 3 GHz, the gain is about 11 dB for the center beams (with weighting by the element factor). For a linear array (shown in Figure 11–7), the elevation beamwidth is about 85°, but can be reduced by ganging antenna elements in this direction. The array elements may be either linearly or circularly polarized.

Assume for the sake of this discussion, that the array operates in the transmit mode. If a signal is injected into Port 2L, several bifurcations occur in the labyrinth, equal ouputs occur at the eight antenna elements, and no energy emerges from the lower ports. Therefore, each signal input from the bottom input ports utilizes the full antenna aperture to form discrete beams (in different directions). Since the array is uniformly illuminated, each antenna beam pattern is of the shape $\sin X/X$, which gives the

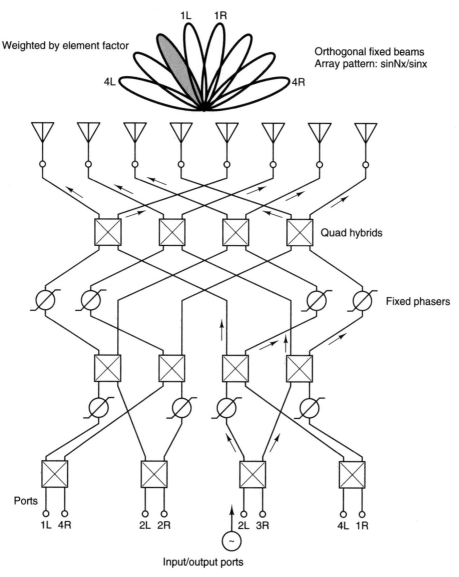

Figure 11-7 A Butler matrix labyrinth producing multiple spot beams, which may be switchable or fixed.

narrowest beam, greatest gain, and a first sidelobe level of −13.2 dB. If all the ports are driven simultaneously, they will produce similar patterns at various azimuth directions. The direction of the beams is dictated by the separation of the antenna elements.

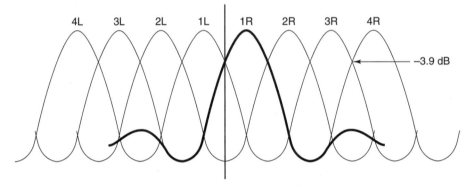

The beams can be decoupled only if the array factor is orthogonal in space. This means that if we multiply one pattern by the other, and integrate the product over all space, the result will be zero.

Since the Butler array is uniformly illuminated, the pattern shape is sinx/x, the patterns are spaced to be orthogonal, and crossover is 4 dB down, with sidelobes down 13.2 dB. Tapering the illumination function, to improve the sidelobe level, will jeopardize this orthogonality.

- Pattern nulls occur at the peaks of the beams.
- There is no element factor weighting shown.

Figure 11–8 Orthogonal beams for an eight-element Butler array.

A typical array factor for an eight-element array is shown in Figure 11–8. Several interesting observations can be made:

- The beams are equally spaced and the peaks are located at the nulls of the other beams.
- Since the array is uniformly illuminated, this gives the smallest beamwidth possible and maximum gain. This follows from array theory. The first sidelobes are down 13.2 dB. Each array pattern has the shape $\sin X/X$, and the array of beams generated is of the form $\sin NX/\sin X$.
- There is beam scalloping with crossover of the beams occurring at the −3.9 dB level.

Configurations

- The beams are orthogonal and outputs are therefore isolated from each other. Orthogonality also implies that the network is lossless save the insertion loss. The latter is kept small by judicious circuit design (<1 dB).
- There is no boresight beam. The Butler matrix, using quadrature hybrids, does not produce a boresight beam. It can be produced if the quadrature hybrids are replaced by hybrid rings, or 0–180° hybrids).

Even though not shown in Figure 11–8, there is a weighting of the antenna patterns by the element factor. This is depicted in Figure 11–9.

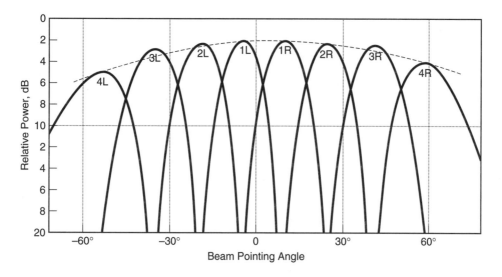

- Major beam patterns of an eight-element array showing the beam distention and gain fall-off as the beams are deployed off boresight.
- The envelope of the peaks follows the antenna element pattern, as we would expect from the principle of pattern multiplication.*

*Array factor times the element factor.

Figure 11–9 Switched/fixed beam array pattern.

11.4 Configurations

Several configurations are possible in the implementation of a switched-beam smart antenna. Figure 11–10 shows a 120° sector covered by four directional beams. The beamwidths are about 30°. The beams are formed by a four-element array and driven by a beam-forming network. A popular beam-forming network is generated by

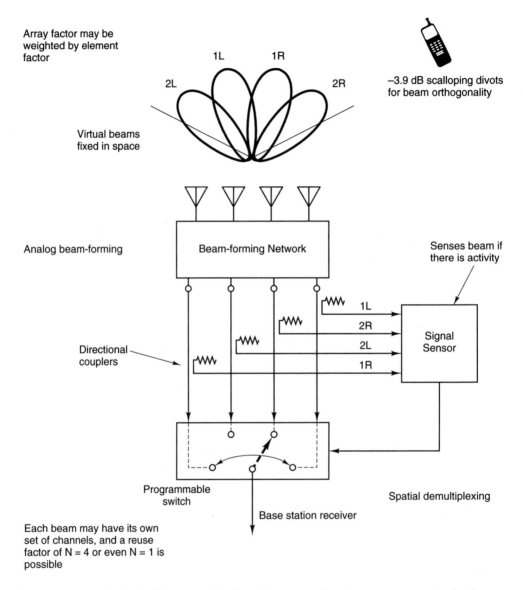

Figure 11–10 Switched-beam technology is a sequel to three-sector sectorization.

the Butler matrix described previously. The directional couplers sample the incoming energy processed by the signal sampler, which in turn programs the switch to the location of the incoming signal from the mobile. This information is passed on to the receiver. For the position of the handset shown, the output will be from Beam $2R$.

Figure 11–11 shows a base station smart antenna which transmits and receives. Each beam is driven by its respective power amplifier shown on the right. The diplexers, of course, are used to transmit and receive at different frequencies (downlink and uplink).

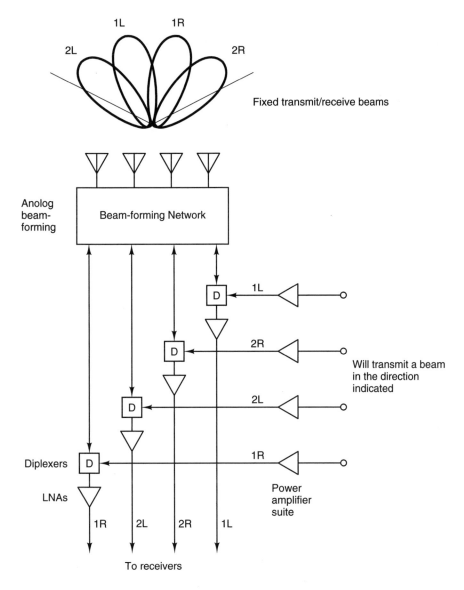

Figure 11–11 Base station system generating four beams in a 120° cell sector.

A COMMENTARY

Occasionally one observes in the literature the use of the word "duplexer". This term is confusing to me. Actually the definition of the word "duplexer" originated in the World War II era and applied to radar applications. That is, a duplexer was used to refer to a switch which allows a radar to transmit and receive at the same frequency, but not at the same time. The diplexer (in this application $f_X \neq f_R$) acts as the vehicle to transmit and receive at different frequencies, but at the same time. Duplexing is the operation, diplexer is the device.

Low-noise amplifiers are used after the diplexers. These, in turn, drive their respective receivers.

Notice that in both of these configurations, processing is done in the analog domain. This may also be executed using digital techniques.

A digital implementation of a switched-beam antenna is depicted in Figure 11–12. The RF signal is heterodyned down to IF. It is then basebanded and converted to digital in the analog-to-digital converter. It is further processed by the digital receiver and then routed to the digital beam-forming network. It is then multiplexed and passed on to the utilization networks.

11.5 Conclusions

Some of the benefits accrued by the use of smart antennas are shown in Table 11–1. The benefits are manifold, and we can expect greater use of these concepts, initially as an accessory to existing omni-directional systems, and in time to supplant them completely. However, this may not come to fruition in rural areas where traffic is light and may not justify the added expense and complexity.

Smart antennas should also lend themselves to all world standards: AMPS, D-AMPS, TDMA, or CDMA. It is my opinion that switched-beam smart antenna systems are more than adequate to satisfy the needs of cellular systems. The ones which have already been implemented have proven to be quite successful; however, they are costly to retrofit.

The other contender for smart antennas is adaptive arrays. This is still in the exploratory stage as far as commercial applications are concerned. The military has been working in this area for many years, especially in the radar sector.

Conclusions

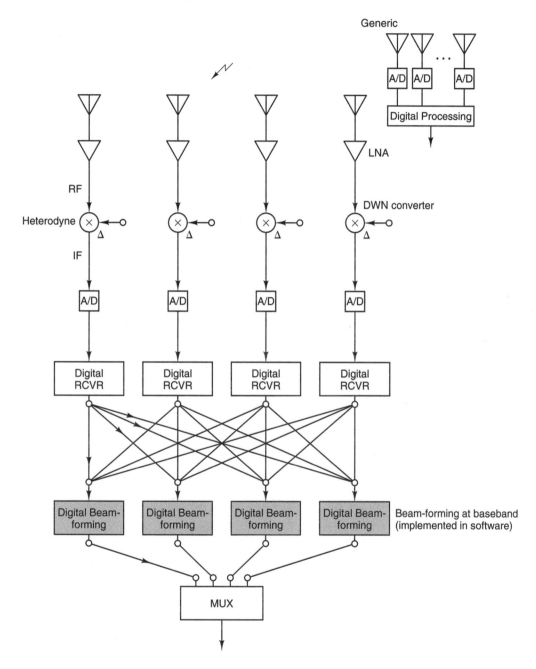

Figure 11–12 Digital implementation of a switched-beam antenna.

Table 11–1 Attributes and Downsides of Smart Antennas

Benefits
- Increased C/I since fewer interferers are in the directional beam(s).
- Increased C/I permits smaller reuse distances ($N : 7 \rightarrow 4 \rightarrow 1$).
- Reduces multipath fading since fewer scatterers are in the beams (multipath may be more deleterious than small C/I).
- In base station transmit mode, greater EIRP can illuminate the mobile.
- On mobile uplink, greater base station antenna gain eases the power requirements of the mobile and increases battery life.
- Can be added to the existing omni infrastructure without loss of service.
- Narrow beamwidth systems may also provide direction of arrival (DOA).
- A 12-beam system can detect signals that are 1/10 the strength of the weakest signal detected by an omni antenna system, and 1/4 the strength of the weakest detected by a three-sector antenna system.
- Switched-beams (in lieu of adaptive) may be good enough for most applications.

On the downside
- Intra-cell handover is required (like in present sectorization), especially if the mobile path is circumferential path.
- More complicated than omni system.
- Directional beams have smaller trunking efficiency.
- Tower may need strengthening or full replacement (expensive) due to added wind loading.

11.6 References

[1] J. Butler and R. Lowe, "Beam Forming Matrix Simplifies Design of Electronically Scanned Antennas," *Electronic Design*, April 12, 1961.

[2] J. P. Shelton and K. S. Kelleher, "Multiple Beams From Linear Arrays," *IRE Trans. on Ant. and Prop.*, March 1961.

[3] J. L. Allen, "A Theoretical Limitation on the Formation of Lossless Multiple Beams in Linear Arrays," *IRE Trans. on Ant. and Prop.*, July 1961.

[4] W. P. Delaney, "An RF Multiple Beam-Forming Technique," *IRE Trans. on Military Electronics*, April 1962.

[5] L. Joseph and W. K. Saunders, "A Theorem on Lossy Nonreciprocal N-Port Junctions," *IRE Letters*, April 1959.

[6] B. Pattan, "The Butler Matrix," *OET Technical Report*, June 1996.

[7] T. Macnamara, "Simplified Design Procedures or Butler Matrices Incorporating 90° Hybrids or 180° Hybrids," *IEE Proc.* (British), Part H, February 1987.

[8] B. Pattan, "Terrestrial-Based Wireless Communications," *OET Internal Technical Report*, 1996.

[9] H. Steyskal, "Digital Beamforming at Rome Laboratory," *Microwave Journal*, February 1996.

CHAPTER 12

Deterministic Signals, Random Noise, and Coherent Noise (Pseudo) Combining in an Array Antenna

12.1 Introduction

On March 6, 1996, a tutorial on smart antennas was given at the FCC. The speaker gave a non-mathematical, but nevertheless interesting, presentation on smart antennas. On one slide (Figure 12–1), he presented a superficial analysis on how signals are combined in an array system. However, it was very sketchy and left out the subtleties underlining the mechanism of performance. We will elucidate on this analysis and attempt to make this concept clearer. We will further discuss other aspects of array theory in this chapter.

Maximum Uplink Gain

SNR improves by factor equal to number of antennas, m

$$\frac{\text{recv'd signal}}{\text{noise}} = \frac{s + \cdots + s}{n_1 + \cdots + n_m}$$

Hence,

$$\text{output SNR} = \frac{(ms)^2}{(m\sigma)^2} = m\frac{s^2}{\sigma^2} = m \times \text{single-antenna SNR}$$

Figure 12–1 Coherent combination of signals in an array antenna and resultant improvement of output signal-to-noise ratio.

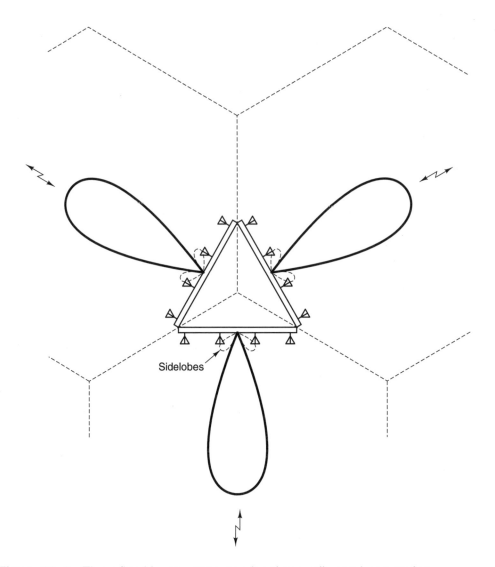

Figure 12–2 Three fixed-beam arrays serving three cells at a base station.

Single beams emanating from the three surfaces of a base station tower are shown in Figure 12–2. Each beam is formed by an array of antenna elements. Each array can be up to one meter long, and could be applied to an existing omni-directional system.

Assume the array consists of four elements separated by 25 cm. At the operating frequency of 800 MHz, the separation is sufficiently close to prevent the onset of grating lobes. This is established from a fundamental relationship in array theory.

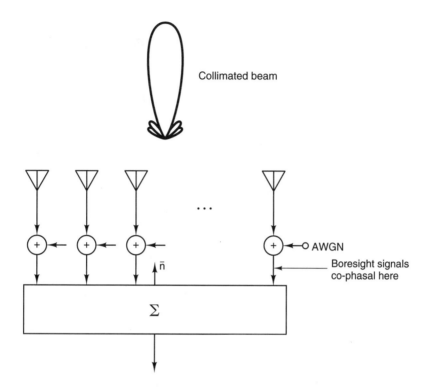

Figure 12–3 Generalized fixed-beam array.

$$(d/\lambda) = [(1 - (1/N)]/1 + \sin\theta \qquad (12.1)$$

where N is the number of elements, d is the element separation, θ is the scan angle (a fixed beam, therefore $\theta = 0°$), and λ is the operating wavelength (assume $f = 800$ MHz). Therefore $d/\lambda = 0.75$, or $d = 37.5(0.75) = 28$ cm.

Because the separation is less than the 25 cm cited above, a grating lobe will not appear in communication space.

For operation at 1.8 GHz, the number of elements typically may be eight. This provides an element separation of about 14 cm. This separation will prevent the advent of grating lobes into communication space.

12.2 Coherent Signals

An array satisfying azimuthal directional requirements is shown in Figure 12–3. This is a linear array and will display a wide beam in elevation. We will not be concerned with the elevation plane in this discussion, however. The antenna elements are shown symbolically, but may be any form of wire antenna at these frequencies.

Because this is a fixed-boresight beam, all deterministic signals in the element arms resulting from the illumination of the aperture by the incident signal from a mobile will be equal and co-phasal. Because the signals are equal and co-phasal, we have

$$e_1 \angle \phi = e_2 \angle \phi = e_3 \angle \phi = \ldots = e_n \angle \phi \tag{12.2}$$

That is, signal voltages at the input to the beam-forming network.

In addition, each element channel will always have the ubiquitous receiver thermal noise, which is given by AWGN as shown.

The summer (Σ) serves as the beam-forming network (BFN). It processes the coherent signal and internal thermal noise differently.

There is an important concept in signal theory which states that while adding deterministic or coherent signals, they are additive on a voltage basis. The phase associated with these signals is instrumental in combining the signals. Random noise, on the other hand, is added power-wise.

A four-element beam-forming network is shown in Figure 12–4. As stated above, the signals at the input ports of the element channels are equal and co-phasal. The combining junction for e_1 and e_2 may be a matched combiner, like a magic tee or reactive combiner. In the matched unit combiner (Junction Ⓐ), the input ports are isolated from each other. If an imbalance occurs at the input ports, the energy is dumped into a matched termination. The reactive combiner unit can cause a mismatch if the incident signals are not equal, but there will be no problem if a match is maintained.

At Junction Ⓐ, the two voltage signals are combined as

$$(e_{s1}/\sqrt{2}) + (e_{s2}/\sqrt{2}) \tag{12.3}$$

At Junction Ⓑ, the outputs are also combined as

$$(e_{s3})/\sqrt{2} + (e_{s4}/\sqrt{2}) \tag{12.4}$$

At Junction Ⓒ, the voltage output of Junction Ⓐ and Ⓑ are combined as

$$e_{so} = \{[e_{s1}/\sqrt{2} + e_{s2}/\sqrt{2}] + [e_{s3}/\sqrt{2} + e_{s4}/\sqrt{2}]\}\sqrt{2} \tag{12.5}$$

The power output of Junction Ⓒ is therefore

$$P_{so} = (e_{so})^2 = (1/N)\left(\sum_{}^{N} e_{si}\right)^2 = (1/N)(Ne_s)^2 = Ne_s^2 = 4e_s^2 \tag{12.6}$$

Coherent Signals

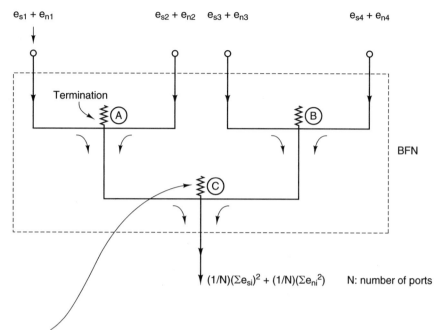

Figure 12–4 Combining deterministic signals and random signals in a corporate combiner.

Therefore, the power output is $N = 4$ times the output from a single antenna channel. Coherent signals have been combined by adding the signal voltages, e_s.

The non-coherent random noise from the four element channels is independent and adds as random noise power in the BFN. At Junction Ⓐ output, we therefore have noise power of

$$P_{nA} = (e_{n1}/\sqrt{2})^2 + (e_{n2}/\sqrt{2})^2 = P_{n1}/2 + P_{n2}/2 \qquad (12.7)$$

at Junction Ⓑ output, we have

$$P_{nB} = (e_{n3}/\sqrt{2})^2 + (e_{n4}/\sqrt{2})^2 = P_{n3}/2 + P_{n4}/2 \qquad (12.8)$$

and at the Junction Ⓒ output, we have

$$P_{nC} = (e_{n1})^2/4 + (e_{n2})^2/4 + (e_{n3})^2/4 + (e_{n4})^2/4 = e_n^2$$

$$\text{or } P_{nC} = (1/N)\sum_{}^{N} e_{ni}^2 = e_n^2 \text{ output noise power} \qquad (12.9)$$

The S/N at the output of the beam-forming network from Equations (12.6) and (12.9) is therefore

$$(S/N) = (P_{so})/P_{no} = 4e_s^2/e_n^2 \qquad (12.10)$$

which is increased by a factor of 4 or generalized to N, the number of antenna elements.

12.3 Coherent Noise

In the above analysis, we assumed that the channel noises are white Gaussian noise. Furthermore, they are independent. They are therefore processed as incoherent noise in the array. However, noise which enters the array from external sources is processed as "coherent noise". The reason for this is that noises which stem from a single source and divide and then combine are correlated. The noise in each channel must therefore be treated as signal-like.

The only way by which the noise can be uncorrelated is when the following correlation relationship is satisfied. The autocorrelation function of a function $f(t)$ is defined as the time average of $f(t)f(t+\tau)$. It is a function of the time interval, τ, and of the function $f(t)$.

$$(\tau) = \overline{f(t)f(t+\tau)} = \lim_{T \to \infty} (1/T) \int_{-T/2}^{T/2} f(t)f(t-\tau)d\tau \qquad (12.11)$$

where τ is the correlation time established by the line lengths in the arrays and any difference in time delay of the signals entering the array. For example, the spatial time differential manifested externally to the array is depicted in Figure 12–5.

For example, if the entering noise were purely white, it would only be correlated with itself, and even an infinitesimal time delay between the two antenna elements would produce independent noise sources. The correlation of white noise is impulsive at $\tau = 0$. This follows from the Weiner-Khintchine Theorem, since the autocorrelation function and signal power spectrum are Fourier transform pairs:

$$\mathfrak{I}[\theta(\tau)] = \Phi(\omega) \qquad \text{power spectrum} \qquad (12.12)$$

$$\mathfrak{I}^{-1}[\Phi(\omega)] = \phi(\tau) \qquad \text{autocorrelation function} \qquad (12.13)$$

Coherent Noise

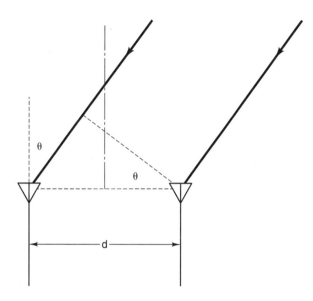

The time difference, Δτ, for the signal to reach the two adjacent elements of the array is related to the angle off boresight from which the signal is received.

Therefore,
 Δτ = (d/c) sin θ
 where d: element separation.

Figure 12–5 Time delay, τ, between two antenna elements.

If the noise spectrum were "brick wall" bounded, as shown in Figure 12–6(a), the autocorrelation function would assume the form sinx/x as shown in Figure 12–6(a). Notice that values of $\tau = k/2B$ ($k = 1, 2, 3, ..., n$) would produce uncorrelated noise output from the channels, even if driven by a single noise source. Note that the differential path length is given as $\Delta l = c\tau$. However, practically a "brick wall" type of spectrum is not realizable and the spectrum skirt is more gradual. An example of this latter spectrum is shown in Figure 12–6(b). Thus, no value of τ will produce uncorrelated noise at the output for spatial noise entry.

In summary, the noise in the main beam of the array must be treated as signal-like, and the beam-forming network does not help to mitigate this "signal".

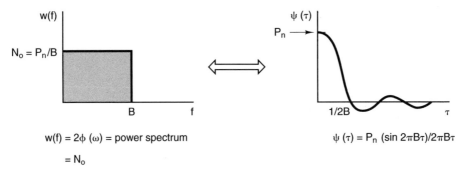

(a) Band-limited noise indicating the discrete delays, "τ", for which the channels' outputs are independent.

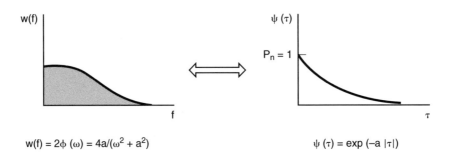

(b) Markoffian noise (RC noise) indicating that the channels' outputs cannot be independent for finite values of "τ".

Figure 12–6 Power spectra and associated correlation functions.

12.4 An Adaptive Array in a Quiescent Signal Field

Consider an N element linear array with elements spaced d distance apart as shown in Figure 12–7. A signal is incident on the array from an angle, θ_s, off boresight. In each element, there is the ubiquitous thermal noise indicated by the variance, θ^2. The noises in each element are independent. In the benign state (no incident interference), the array is electronically steered by the element-weighting terms, which steer the beam in the desired direction. The output of the phasers (weights) are thus co-phasal, and the signals are summed coherently to give the maximum output S/N ratio.

In the adaptive array, the phasers are weights which negate the incoming signal phase so that the signals are optimally weighted to give the maximum output to the bona fide signal; and nulls otherwise if interference is present.

An Adaptive Array in a Quiescent Signal Field   235

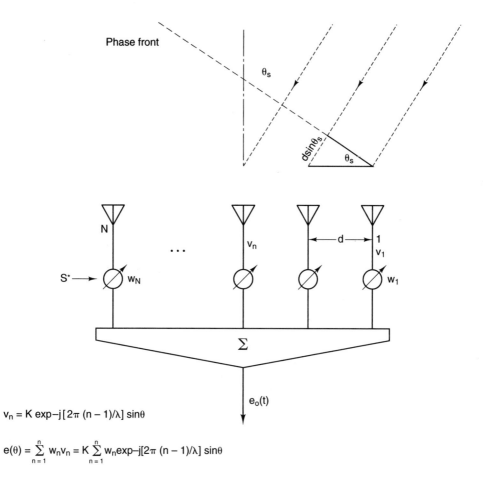

Figure 12–7 Receiving scenario in an adaptive array.

A signal vector whose components are the complex signals in the columns can be given by

$$S = [s_1(t), s_2(t), ..., s_N(t)]^t \tag{12.14}$$

where t is the transpose operation. The complex weights (which may consist of amplitude control for aperture tailoring and phase control for steering) can be put in terms of a complex weight vector

$$W = [w_1, w_2, ..., w_N] \tag{12.15}$$

The array output therefore becomes

$$e_o(t) = W^t S \quad (12.16)$$

To maximize the array output, it is necessary to specify a vector, S (as shown above), or a vector of voltage induced in the array by a wave driving the array from the desired direction, that is, the space vector which describes the relative phase and amplitude of the desired signal at each element of the array. This, coupled with the covariance matrix, produces optimum weighting. The weight vector is therefore optimized if

$$W = M^{-1} S^* \quad (12.17)$$

where M^{-1} is the inverse of the covariance matrix, S^* is the conjugate of the desired signal space vector, designated previously by S. This conjugate vector therefore becomes:

$$S^* = \begin{bmatrix} a_1 \\ a_2 \exp-j[(2\pi d/\lambda)\sin\theta_s] \\ a_3 \exp-j[(4\pi d/\lambda)\sin\theta_s] \\ \cdot \\ \cdot \\ \cdot \\ a_N \exp-j[(N-1)(2\pi d/\lambda)]\sin\theta_s \end{bmatrix} 1 \quad (12.18)$$

where a_n is amplitude taper (if used) and ϕ is $(2\pi/\lambda)d(N-1)\sin\theta_s$.

The phase of a signal arriving from a distant source displaced from boresight by θ_s will have this relative phase shift between elements. The phase shift increases linearly across the array. The complex conjugate phase shifts, given by Equation (12.18), are introduced in the weights. This will therefore permit coherent summation of the individual antenna element's output. Note the incoming signal vector is the conjugate of the steering vector, S.

As Equation (12.17) shows, to find the optimized weights, it is necessary to find the inverse of the covariance matrix. In the next section, we will consider this exercise.

---

1. The Applebaum algorithm maximizes a generalized S/N output of the array, and the DOA of the signal is known (usually in radar and satellite applications), hence, the steering vector injection. In the Widrow algorithm, the signal DOA is not known apriori (the scenario in communications applications), but an internally-generated reference signal is available. The S term in this case is the cross-correlation between the individual input signal samples and the error signal formed by the output-reference signal.

An Adaptive Array in a Quiescent Signal Field

The generalized form of the matrix M is given as:

$$M = \begin{bmatrix} m_{ll} & \cdots & m_{ln} \\ \cdot & & \cdot \\ \cdot & & \cdot \\ \cdot & & \cdot \\ m_{nl} & \cdots & m_{nn} \end{bmatrix} \qquad (12.19)$$

This is the covariance matrix (correlation matrix). Elements of this matrix denote the correlation between various antenna elements. For example, m_{ij} denotes the correlation between the ith and j^{th} element of the array. The noise voltages between channels are independent of each other.

Its inverse is given as

$$M^{-1} = (1/\det M)[c_{ij}]^t = (1/\det M) \begin{bmatrix} c_{ll} & \cdots & c_{nl} \\ \cdot & & \cdot \\ \cdot & & \cdot \\ \cdot & & \cdot \\ c_{ln} & \cdots & c_{nn} \end{bmatrix}^t \qquad (12.20)$$

where c_{ij} is the cofactor of each of the original elements, m_{ij} in M, which is defined as the signal minors of the matrix when the ith row and j^{th} column are removed det M is the determinant of matrix M and t is the transpose operation (after plugging in the cofactors).

12.4.1 An Example

As an illustration, we can simplify the analysis by assuming that M is a 2×2 matrix. It is a simple matter, but laborious to deal with, a matrix of an order greater than two. We therefore obtain

$$M = \begin{bmatrix} m_{11} & m_{12} \\ m_{21} & m_{22} \end{bmatrix} \qquad M^{-1} = (1/\det M) \begin{bmatrix} c_{11} & c_{12} \\ c_{21} & c_{22} \end{bmatrix} \qquad (12.21)$$

The cofactors of M are

$c_{11} = m_{22}$ (e.g., Row 1 and Column 1 are removed in M)
$c_{12} = -m_{21}$
$c_{21} = -m_{12}$
$c_{22} = m_{11}$ (12.22)

Note all are multiplied by the signed $(-1)^{i+j}$.
The determinant of M is

$$\det M = m_{11}m_{22} - m_{21}m_{12} \tag{12.23}$$

Thus,
$$[c_{i,j}] = \begin{bmatrix} m_{22} & -m_{21} \\ -m_{12} & +m_{11} \end{bmatrix} \tag{12.24}$$

and
$$[c_{i,j}]^t = \begin{bmatrix} m_{22} & -m_{21} \\ -m_{12} & +m_{11} \end{bmatrix} \tag{12.25}$$

Hence, the inverse matrix of matrix M, given by Equation (12.21) is

$$M^{-1} = \begin{bmatrix} m_{22}/m_{11}m_{22} - m_{21}m_{12} & -m_{12}/m_{11}m_{22} - m_{21}m_{12} \\ -m_{21}/m_{11}m_{22} - m_{21}m_{12} & m_{11}/m_{11}m_{22} - m_{21}m_{12} \end{bmatrix} \tag{12.26}$$

The M matrix is defined as a matrix which has the property $[M][M]^{-1} = [I]$, where $[I]$ is the identity matrix where all diagonal elements are unity and all off-diagonal elements are zero. M, therefore, must be a square matrix.

Previously we showed that in the absence of external interference (quiescent field), the off-diagonal elements in the inverse matrix are equal to zero, and the diagonal elements are the independent thermal noise powers.

The last matrix, Equation (12.26), then simplifies to

$$M^{-1} = \begin{bmatrix} \sigma^2/\sigma^4 & 0 \\ 0 & \sigma^2/\sigma^4 \end{bmatrix} \tag{12.27}$$

where $m_{22} = m_{11} = \sigma^2$ and $m_{11}m_{22} = \sigma^4$. Therefore, Equation (12.27) becomes

$$M^{-1} = 1/\sigma^2 \begin{bmatrix} 1 & 0 \\ 0 & 1 \end{bmatrix} = 1/\sigma^2 [I] \tag{12.28}$$

Since $W = M^{-1}S^*$, this can be combined with Equation (12.28) to give

$$W = (1/\sigma^2) S^* \quad (12.29)$$

which states that the optimum weights are the complex conjugate of the desired signal.

In a benign environment (no interference, but thermal noise present), the W vector is merely the deterministic element. Phasers are instrumental in producing co-phasal signals at the input to the summer. This gives the maximum output S/N ratio. The nose of the beam is pointing at the spatial signal (thanks to S^*) coming in at an angle θ_s degrees off boresight.

CHAPTER 13

Adaptive Arrays in Cellular Communications

13.1 Introduction

In a phased array receiving antenna, the signals received by the array elements are generally added at RF to form a receiving beam. In an adaptive array, the phase and amplitude of each element output are controlled by an adaptive network, that is, they use algorithms that iteratively adjust the weighting of the signals at the array elements. The signals are combined to maximize the signal-to-interference-plus-noise ratio (SINR) and to eliminate interfering signals in any direction other than the main beam.

A simplistic form of this array is shown in Figure 13–1. We depict a linear array, but this can be generalized to a planar array, or even a conformal array. The weights are controlled depending on the signal and the noise/interference environment as well as by the system requirements. These weights are complex in that they provide both amplitude and phase information.

The output of each element is $e_1(t)$, $e_2(t)$, $e_3(t)$, ..., $e_N(t)$. The beam (assumed receiving) is formed by multipling each $e(t)$ by a complex weight, w_n, and then summing the resultant signals. Both w_n and e_n are complex quantities in that they are variable in phase and amplitude.

By suitable choice of complex conjugate phase shifts in the weights[1], a beam can be steered to the direction of the desired off-boresight angle, θ_s, which gives a coherent

---

1. Weights are the complex conjugate of the channel transfer characteristic. In general, they have continuous amplitude and phase characteristics.

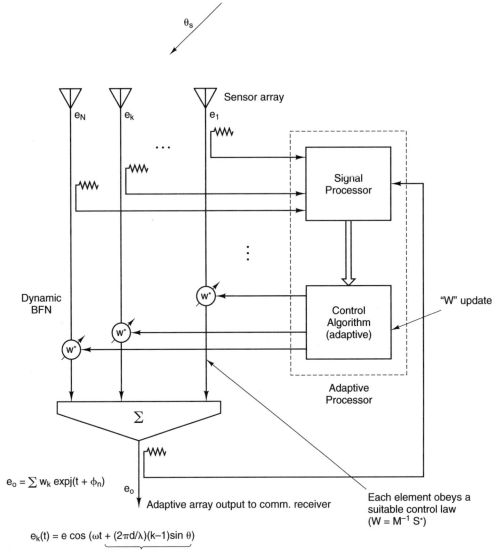

$e_o = \sum w_k \exp j(t + \phi_n)$

Adaptive array output to comm. receiver

Each element obeys a suitable control law $(W = M^{-1} S^*)$

$e_k(t) = e \cos(\omega t + (2\pi d/\lambda)(k-1)\sin\theta)$

ϕ_n: relative phase shift of received signal at k element.

- The array is described as adaptive when the weights are calculated by a closed-loop feedback algorithm. The weights are continuously updated to minimize the power of the interfering signal at the array output and to optimize the signal. The feedback is used to vary the phase and/or amplitude weighting of the signals received in each antenna element.

Figure 13–1 Adaptive array.

summation of the individual inputs and a null in an interfering direction, that is, at an angle other than at θ_s.

In cellular applications, when used at the base stations, they are capable of providing automatic adjustment of the far field pattern to form nulls in the direction of interfering sources while maintaining high gain in the direction of the desired user.

13.2 The Theory

From Figure 13–1, a signal vector whose components are the complex signals on the row of elements is denoted by the matrix

$$E = [e_1(t), e_2(t), e_3(t), ..., e_N(t)]^t \qquad (13.1)$$

in which t denotes the transpose operation. E is multiplied by a complex weight vector of the beam former, as in

$$W = [w_1, w_2, w_3, ...]^t \qquad (13.2)$$

to produce the array output. We therefore have

$$s(t) = [w_1, w_2, w_3, ..., w_N] \begin{bmatrix} e_1(t) \\ e_2(t) \\ e_3(t) \\ \cdot \\ \cdot \\ \cdot \\ e_N(t) \end{bmatrix} \qquad (13.3)$$

consequently, $s(t)$ is the product of a row vector and a column vector. Or, in abridged matrix form,

$$s(t) = W^t E = E^t W \qquad \text{scalar output} \qquad (13.4)$$

The signal output of the array is therefore

$$s(t) = w_1 e_1(t) + w_2 e_2(t) + w_3 e_3(t) + ... + w_N e_N(t)$$

$$= \sum_{m=i}^{N} w_i e_i(t) \qquad (13.5)$$

Returning to Figure 13–1, if we assume that the first element (top) of the array is the phase reference, the relative phase shift of the received signal at the nth element is

$$\phi_n = e[(2\pi(d-1)]/\lambda)\sin\theta_s \qquad (13.6)$$

where θ_s is the direction of the virtual receive beam. The signal at the nth element is

$$e_n(t) = e \exp j(\omega t + \phi_n) \qquad (13.7)$$

Plugging in Equation (13.5) from (13.6) and (13.7) we have

$$s(t) = \sum_{m=1}^{N} ew_m \exp j[\omega t + (2\pi(d-1)/\lambda)\sin\theta_s] \qquad (13.8)$$

Actually, we could relegate the cissoid to limbo, since it has no influence on the analysis.

We observe that if the complex conjugate weights are only conjugate to the signal vectors, there will be a beam formed in the direction of angle θ_s off boresight. It may be of further interest to note that if we introduce an *amplitude* control (real part of complex weight), a taper may be imparted to the aperture producing lower antenna sidelobes.

In the adaptive array, the receive beam is formed by adding the complex weighted outputs of the array elements, as shown in Figure 13–1. Each of the element weights, w_n, is controlled by a separate adaptive loop. The weights are time-varying and programmed to the signals (deterministic and random) in the channels.

The following parallels the development of adaptive arrays given by Applebaum in his classic report [1]: Consider the array shown in Figure 13–1. Assume the signal received is an interferer which is added to the channel's thermal noise. Assume a vector, $\vec{E}$, whose elements are the voltages in the channel induced by the sources of noise. We therefore form a matrix, M, according to the equation

$$M = \mathcal{E}(E^*E^t) \qquad (13.9)$$

that is, the conjugate of the complex vector $\vec{E}$ is multiplied by the transpose of the vector. $\mathcal{E}(...)$ denotes the expectation, or average value of a quantity, in this case, noises. The result is known as the *covariance matrix* of the noise components, and is fundamental to all processor theory. The statistics of the total noise field, interference plus internal noise, is represented by the covariance matrix, M, which contains second central moments[2] (variances) of the noise process. It contains all the information about the interference environment and denotes the correlation between various element noises.

---

2. The mean is the first moment, but the first moments are equal to zero.

We note that there is no correlation between quiescent (channel) noise in different channels, but there is *correlation* between interference noise in the other channels. This has also been referred to as coherent noise. More will be said on this later.

The covariance matrix can assume different values depending on the noise environment. Clearly, a pure thermal environment, in which channel noises are uncorrelated, differs from a noise situation in which noise enters the array. Here, the noises are correlative (we brought this concept to light in the previous chapter).

The *diagonal* terms are a measure of the total noise power in the channels; the *off-diagonal* terms contain information about the direction of arrival of the interfering noise. The intrinsic channel noises are independent of each other, and of the interfering noises. For example, for a four-element array, the covariance matrix is of the following form. This follows from Equation (13.9), or the expected values of the noise sources in the channels.

$$M = \begin{bmatrix} (N+I) & I & I & I \\ I & (N+I) & I & I \\ I & I & (N+I) & I \\ I & I & I & (N+I) \end{bmatrix} \quad (13.10)$$

where σ_n^2 labels the diagonal and σ_I^2 labels the off-diagonal.

The off-diagonal terms also contain information about the direction of arrival of the various interference signals

where $N = \sigma_n^2$ (second moment about the mean): the thermal noise power in each channel.

This is usually considered as additive white Gaussian noise (AWGN), with zero mean and variance equal to σ^2, I is the interfering power.

To find the elements of a 2×2 M matrix is a bit more tractable than finding those for a 4×4 M matrix. This would correspond to the following two-element array:

$$M = \begin{bmatrix} m_{11} & m_{12} \\ m_{21} & m_{22} \end{bmatrix} \quad (13.11)$$

$$m_{11} = \mathcal{E}(e_1^* e_2) = \mathcal{E}(i_1^* i_2) + \mathcal{E}(n_1^* n_1) = I + N = \sigma_i^2 + \sigma_n^2$$
$$m_{12} = \mathcal{E}(e_1^* e_2) = \mathcal{E}(i_1^* i_2) = I \exp(-j\phi)$$
$$m_{21} = \mathcal{E}(e_2^* e_1) = \mathcal{E}(i_2^* i_1) + \mathcal{E}(i_2^* i_1) = I \exp(j\phi)$$
$$m_{22} = \mathcal{E}(e_2^* e_2) = \mathcal{E}(i_2^* i_2) + \mathcal{E}(n_2^* n_2) = I + N \tag{13.12}$$

The i's are interference and the n's are channel noise only. I and N are noise powers from channel noise and interference noise, respectively.

Note that there is independence between channel noise in other channels, correlation of channel noise within the channels, and independence between interference noise in the channel. Similarly, there is dependence between interference between different channels and also within a channel.

From the matrix elements above, we have

$$\mathcal{E}(i_1^* i_1) = \mathcal{E}(i_2^* i_2) = I \text{ and } \mathcal{E}(n_1^* n_1) = \mathcal{E}(n_2^* n_2) = N \tag{13.13}$$

It is interesting to note that when the interference is equal to zero, or $N (=\sigma_n^2) \gg I$, the covarinace matrix is a diagonal matrix.

$$M_q = \begin{bmatrix} \sigma_n^2 & 0 \\ 0 & \sigma_n^2 \end{bmatrix} = \sigma_n^2 I \tag{13.14}$$

where I is an identity matrix of order $n(2)$ and the subscript q indicates that this is a quiescent, or noise-only state.

The diagonal matrix further reveals that the noise conponents are uncorrelated. In general, if there is interference present, the matrix "fills up," as is shown in the matrix, M, above.

To compute the set of adaptive coefficients, or weights, it is necessary to specify the direction in which we wish to steer the main beam. We then define a vector, S, which is the vector of the voltage induced in the array from a desired steering direction. Applebaum's paper shows that the weights which yield the best SNIR for a signal arriving from a desired direction is given by

$$W_{opt} = u M^{-1} S^* \tag{13.15}$$

S is the steering vector, and represents weights which establish the antenna receiving pattern in the absence of interfering sources

where u is an arbitrary constant, and S^* is the set of steering signals matched to the desired signal response (i.e., matched to the angle of the incident signal).

Note that the array weights depend on the external interference represented by the interference covariance matrix, M.

The covariance matrix is inverted and multiplied by the conjugate of the steering vector.[3] This is the optimum set of complex weights, which maximizes the ratio of the main beam gain to the total noise $(N + I)$ power at the output.

Depending on the number of elements, finding the inverse of the covariance matrix in Equation (13.10)(13.15) could be complicated and lengthy. In fact, a computer may be necessary to perform this inversion.

In the matrix where $I = 0$ above, the inverse matrix follows as

$$M^{-1} = \begin{bmatrix} 1/\sigma^2 & & & \bigcirc \\ & 1/\sigma^2 & & \\ & & 1/\sigma^2 & \\ \bigcirc & & & 1/\sigma^2 \end{bmatrix} \quad (13.16)$$

and Equation (13.15) becomes

$$W = \begin{bmatrix} 1/\sigma^2 & & & \bigcirc \\ & 1/\sigma^2 & & \\ & & 1/\sigma^2 & \\ \bigcirc & & & 1/\sigma^2 \end{bmatrix} S^*$$

$$= u(1/\sigma^2)S^* \quad (13.17)$$

Here we multiplied a unity matrix into S^*, and further stated that the optimum weights were the complex conjugate of only the desired signal (i.e., there is no interferece present).

The desired steering vector for this case is when there is no external interference, or a signal in the direction θ_s becomes (conjugated) for a four-element array

---

3. Note the conjugation negates the different phasing of the signals incident on the array coming in at an angle θ_s frm boresight. W may also contain terms to tailor the illumination function.

$$S* \begin{bmatrix} a_1 \\ a_2 \exp j\phi \\ a_3 \exp j2\phi \\ a_4 \exp j3\phi \end{bmatrix} = \begin{bmatrix} a_1 \\ a_2 \exp j(2\pi d/\lambda)\sin\theta_s \\ a_3 \exp j2(2\pi d/\lambda)\sin\theta_s \\ a_4 \exp j3(2\pi d/\lambda)\sin\theta_s \end{bmatrix} \qquad (13.18)$$

where d is the element separation and θ_s is the scan angle.

For a uniformly weighted aperture, all the a terms shown above are equal, and clearly we have a pattern display of $\sin Nx / \sin X$.[4] This would give a first sidelobe level of 13.2 dB down. In practice though, the a's are generally tailored to reduce the sidelobes (with a small loss in on-axis gain and a small reduction in output *S/N* ratio).

For a more generalized case where interference is present, the covariance matrix is no longer a diagaonal matrix since some of the noise components are now correlated. This was shown previously in the *M* matrix. In this case, the weighting vector takes on a new set of values to cope with the interference.

Three regimes of weighting can be manifested:

1. In a benign environment, in which only the channel thermal noise is present. Noises here are assumed uncorrelated (which is also the practical case). For uniform illumination, the steering terms are as given in classical array theory and the weights are pure imaginary (phasers in each channel).
2. When amplitude weighting is produced across the aperture to provide a better antenna pattern. This is still a benign environment. The steering vectors are complex (real + imaginary term).
3. When there is external interference entering the array. Here, the weighting is performed to reduce the interference coming from other than the desired beam position. The interfering signal in the channels must be correlated for the complex weight to process and produce a signal at the summer. There are factors which can produce decorrelation and reduce the depth of the null(s).

13.3 Simulation Results

It is interesting to note what happens in an adaptive array in the presence of a single interferer. The optimum weighting factor can be shown to consist of two compo-

---

4. A uniformly weighted array aperature gives the maximum S/N when channel noises are equal and uncorrelated. If interference is incident on the array, different weighting is required to optimize the output S/(N+I) ratio.

Simulation Results

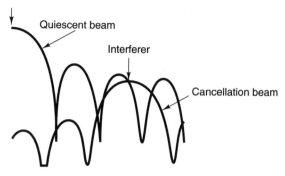

(a) Quiescent and cancellation beams (top); adapted composite pattern (bottom).

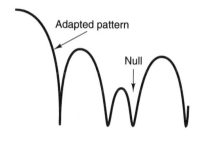

(b)

(c)

Figure 13–2 Sketch of the pattern of an adaptively controlled array. Resultant pattern is the difference between the quiescent and the cancellation pattern.

nents. This has been shown by Gabriel [5], and is tantamount to dividing the covariance matrix into the quiescent, receiver-only noise matrix, interfering source matrix, and as a result, two antenna patterns. The resultant patterns are shown in Figure 13–2. For the quiescent situation, in which there is no interference, the pattern is the non-adaptive array beam. This produces the $\sin x/x$ beam shown. The other matrix produces a similar beam, but it is pointed at the interferer. The latter has been referred to as the cancellation beam.

Both of these beams are depicted in Figure 13–2(a). When the cancellation beam is subtracted from the quiescent beam, the composite result is shown in the lower part of Figure 13–2(a).

Another representation is shown in Figure 13–2(b).

13.4 Conclusions

The level of interference rejection and the ability to maintain a reasonably shaped beam are the primary measures of performance of an adaptive array. Adaptive arrays can be configured to sense the external noise field and drive the element weights, w_i, to optimum values. An adaptive system will automatically place a null(s) in the antenna pattern at the angle of the discrete cellular interferers.

In the case of an adaptive array, in which there are no external interferers present, the array performs as a classical non-adaptive array in which the weights contain the phase gradient to steer the beam, as well as the amplitude weights for tailoring the illumination function. In the channels, the ubiquitous thermal noises are independent and the array output will manifest the greatest possible S/N ratio.

13.5 References

[1] S. P. Applebaum, "Adaptive Arrays," Syracuse Univ. Research Corp., Report SURC TR 66-001, March 1975.

[2] R. A. Monzingo and T. W. Miller, *Introduction to Adaptive Arrays*, John Wiley and Sons, New York.

[3] Special Issue on Adaptive Antennas, *IEEE Ant. and Prop. Transactions*, September 1976.

[4] B. D. Carlson, et al., "An Ultralow Sidelobe Adaptive Array Antenna," *The Lincoln Laboratory Journal*, Vol 3, #2, Summer 1990.

[5] W. F. Gabriel, "Adaptive Arrays-An Introduction," *Proc. IEEE*, February 1976.

References

[6] D. J. Tottieri, *Principles of Military Communications Systems*, Artech House, Dedham, MA, 1981.

[7] L. C. Godara, "Application of Antenna Arrays to Mobile Communications," *Proc. IEEE*, August 1997.

[8] F. Ayres, Jr., *Matrices*, Schaum Publishing Co., New York, 1962.

[9] H. Eves, *Elementary Matrix Theory*, Dover Pub. Inc., New York, 1966.

[10] S. Perlis, *Theory of Matrices*, Addison-Wesley Inc., Cambridge, MA, 1952.

CHAPTER 14

Summary – Smart Antennas in Cellular Communications

14.1 Introduction

Dr. Andrew Viterbi has said that spatial filtering is the next step in improving the capacity of cellular systems. Spatial filtering will complement the other technologies which have been exploited as well, such as cell splitting, innovative signal formats, source signal compression, and others.

Basically, spatial filtering uses the properties of directional antennas (arrays) to maximize response in the direction of the desired signal and minimize response in the direction of the interfering signals. Clearly the latter is attenuated by residing in the sidelobes of the directional antenna with the interfering level being down generally greater than 10 dB.

Spatial filtering is the next step in the evolution of cellular systems. The boxed material, Evolution of Cellular Systems, shows this from the first generation, commencing in the early 1980s, to the more recent era of smart antennas. These smart antenna systems are being developed, and there are companies which are presently marketing such systems as applique to existing cellular systems. Two problems which have been manifested by attaching a smart antenna, which is usually an array of antenna elements and support circuitry, to an existing base station tower are the additional wind loading and reduced reliability. Nevertheless, these problems are apparently being resolved.

As has been indicated in previous chapters, there are basically two types of smart antennas. These are indicated in the boxed material entitled Smart Antenna Genres. In the fixed/switched-beam concept, the directional beams deployed at a base station may

Evolution of Cellular Systems

- First Generation
  - Analog Modulation (AMPS-FM).
  - Omni-Directional Antennas at Base Station.
  - Seven-Cell Clustering.
- Second Generation
  - Cell Splitting.
  - Cell Sectorization.
  - Digital Modulation.
  - Less than $N = 7$ Cell Clustering (Reduced Reuse Distance).
- Third Generation
  - Emergence of SDMA via Switched Directional Beams and Adaptive Nulling Array Antennas.

Smart Antenna Genres

- Fixed/Switched-Beam: Uses a beam-forming network (BFN) to provide fixed directional beams with output ports (INPUT) corresponding to fixed-beam positions. Output may be sampled simultaneously or sequentially switched to ascertain beam activity. Hand-off between beams is required as users undergo tangential motion.
- Adaptive Array: An antenna array that can dynamically modify its antenna pattern structure to place a directional beam in the direction of a bona fide signal (thus maximizing the output $S/N+I$) and nulls in the direction of the interferers. Signal location must be known apriori (Applebaum-Howells genre), or not known; however, a pilot replica is required (Widrow genre array)

be fixed in space or switchable. In the fixed-beam case, a switching matrix is used to address a particular user. An example of a fixed/switched beam is depicted in Figure 14–1. A Butler beam-forming labyrinth is shown at the left of the figure. Notice that the

Introduction

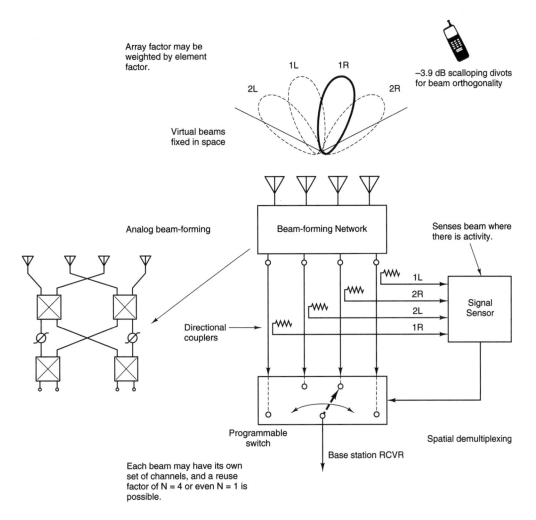

Figure 14–1 Fixed- or switched-beam (shown) array.

beams are shown in the azimuthal direction, and in elevation, the beam is usually broader than in the azimuth direction. This is understandable since a user may come in from several elevation angles.

In a more elaborate adaptive array (see subsequent sections in this chapter), a beam is placed in the direction of cellular activity and nulls are placed in the direction of potential interferers. Again the interferer(s) are placed in the null(s) of the antenna pattern.

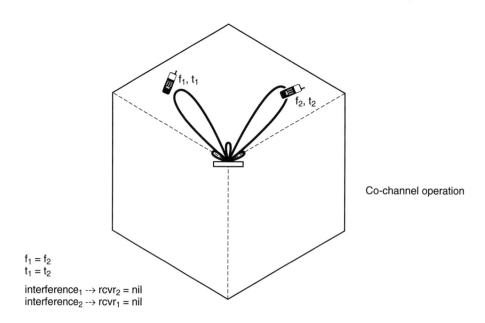

Figure 14–2 Frequency reuse within a cell.

One of the benefits which accrue from the use of smart beams is that users residing in different beams but in the same cell are able to reuse intra-cell frequency. A simplified display of this performance is indicated in Figure 14–2.

Some of the properties of both switched-beam and adaptive-array technologies are listed in the boxed material on page 257. Generally, adaptive arrays are more complicated than switched-beam types because they are more processing intensive.

14.2 Adaptive Array Genre

Different types of adaptive arrays considered for wireless applications include the Applebaum and Widrow configurations. These are shown in their most elementary form in Figure 14–3. In the Applebaum network, the weights in the adaptive loop are adjusted by the feedback output. In the Widrow scheme, a replica of the desired signal is available and compared with the output to generate an error signal. More detailed configurations are depicted in Figure 14–4.

One interesting subtlety is observed in an adaptive array implementation using *analog* weighting in the RF channels. The channels cannot cope with more than *one* user in the cell. The reason is that two users are at different bearings and would require different RF weightings (at the same time). Therefore, in the analog domain, only one

Adaptive Array Genre

Properties of Switched-Beam/Adaptive-Array Technologies

- Switched-Beam Technology
  - High gain, narrow azimuth beamwidths.
  - Butler Matrix Array Technology (or other BFNs).
  - Multiple fixed-direction beams.
  - Algorithms for beam selection
  - Hand-off between beams required (tangential).
  - Interference is beam dependent.
- Adaptive-Array Technology
  - Antenna beam adaptive to signal direction; null on interference.
  - No hand-off between beams (TRKS signal location).
  - Constant antenna gain for all users since beam dynamically "searchlights" signal source.
  - May have greater capacity increase than switched beam.
  - More intensive processing using DSP.

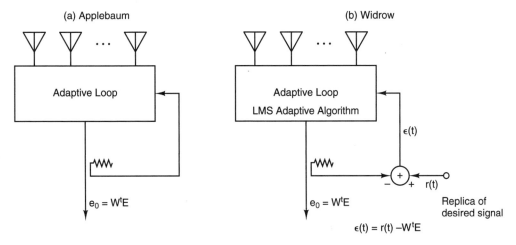

Figure 14-3 The basic differences between Applebaum and Widrow adaptive arrays.

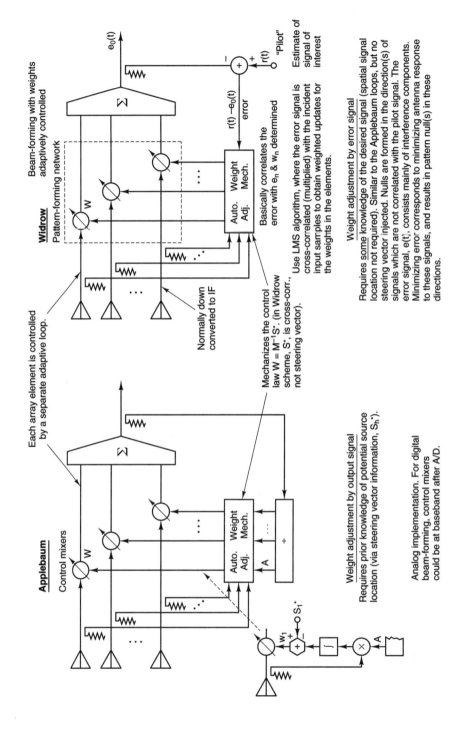

Figure 14-4 Two basic adaptive-array configurations.

user can be accommodated at a time. However, if Digital Signal Processing (DSP) is performed after the RF mixing process, different weights can be assigned to different users located at different bearings.

Even though the Applebaum concept appears attractive in several respects, it would have limited application in wireless since user location is required apriori. That is, a steering vector is provided to the adaptive loop. Therefore, it would find its greatest applications in radar or satellite applications since the "users" are already known. This also applies to the previously discussed sidelobe canceller, which is a derivative of, or precursor to, the Applebaum solution.

The boxed material, Algorithms Used for Controlling Adaptive Arrays, indicates the algorithms used for controlling adaptive arrays.

Algorithms Used for Controlling Adaptive Arrays

- **Applebaum Algorithm:** Adaptive-array technique for maximizing $S / N + I$ at the output. The signal bearing (steering vector) *is known* and the algorithm seeks to maximize the output.
- **Widrow Algorithm:** Minimizes the mean square error signal which is the difference between the array output and a reference signal. The weights are adjusted so the mean square value of $\varepsilon(T)$ between the array output and the reference signal is minimized. Will form and /or steer a directional beam in an unknown user's direction, and place nulls on all interferers, where the signals do not correlate (LMS algorithm). n.b.: These are closely related in that they both supply optimizing feedback signals to the antenna elements. The array output is the weighted sum of the array elements. Attempt to acheive steady-state weights (optimum) yields $W_{opt} = M^{-1} S^*$, where W is the column vector of weights, M^{-1} is the inverse of the covariance matrix, and S^* is the steering vector in the Applebaum scheme or the cross-correlation vector between the received signal and the reference signal in the Widrow implementation.

If no interference:
$M = O^2 I$
$W = (1 / O^2) S^*$
The optimum weights are the complex conjugate of the desired signal.

14.3 Where Are Smart Antennas Going?

Lucent Technologies (actively engaged in adaptive-array technology) suggests that adaptive arrays of the Widrow genre will be the smart antennas of the future.[1] Switched-beam concepts will not reach the capacity of adaptive arrays. Instead, switched-beam concepts will be more applicable in CDMA systems, where the base station is surrounded by interferers.

14.4 Conclusions

Smart antennas will improve performance in several ways:

- Reduced co-channel interference (CCI).
- Reduced multipath interference via increased antenna beam directionality.
- Reduced delay spread[2] because fewer scatterers are illuminated.
- Increase frequency reuse with fewer base stations.
- Increase spectral efficiency or network capacity.
- Increase range in rural areas.
- Higher gain supports higher data rates.
- Improved building penetration.

---

1. ARRAYCOM, Inc. has a system dubbed "INTELLICELL" which uses an adaptive array. The system was described in *very* general terms at the FCC tutorial, but no details were revealed. This writter suggests it may be very similar to the system being developed by Lucent for the IS-136 standard (D-AMPS).
2. Delay spread: Delayed versions of an original signal are produced when a transmitted signal is subjected to multipath propagation in its transit. The effect is called *delay spread*. In a digital system, delay spread causes intersymbol interference (ISI).

APPENDIX A

Gaussian Low-Pass Filter

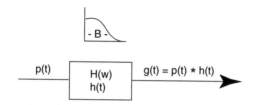

$$H(w) = \tau\sqrt{2\pi} \, \exp\left[-\left(\frac{1}{2}\right)(\tau w)^2\right] \quad \text{Gaussian shape}$$

where τ is constant. Since the system function, $H(w)$, and the impulse response, $h(t)$, are Fourier transform pairs, we obtain

$$h(t) = \mathfrak{I}^{-1}[H(w)] = \left(\frac{1}{2}\pi\right)\int_{-\infty}^{\infty} H(w)\exp(+jwt)dw$$

$$= (1/2\pi)\int_{-\infty}^{\infty} \tau 2\pi\exp\left[-\left(\frac{1}{2}\right)(\tau w)^2\right]\exp(+jwt)dw$$

$$= \exp-\left(\frac{1}{2}\right)\left(\frac{t}{\tau}\right)^2 \quad \text{also Gaussian shape}$$

$$g(t) = p(t)*h(t) = \int_0^T h(t-\tau)p(\tau)d\tau$$

$$G(w) = P(w)(H(w))$$

$$\text{if } P(t) = \delta(t)$$

$$P(w) = \mathfrak{I}[\delta(t)] = 1$$

$$\therefore G(w) = H(w)$$

$$g(t) = \mathfrak{I}^{-1}H(w) = h(t)$$

APPENDIX B

Scattering Matrix of the Quadrature Hybrid

The general scattering matrix for a four-port junction is:

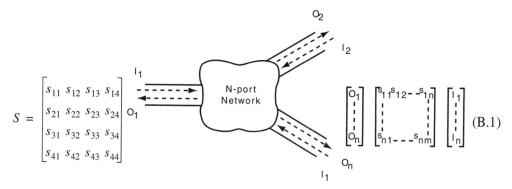

$$S = \begin{bmatrix} s_{11} & s_{12} & s_{13} & s_{14} \\ s_{21} & s_{22} & s_{23} & s_{24} \\ s_{31} & s_{32} & s_{33} & s_{34} \\ s_{41} & s_{42} & s_{43} & s_{44} \end{bmatrix} \quad \begin{bmatrix} O_1 \\ \vdots \\ O_n \end{bmatrix} \begin{bmatrix} s_{11} & s_{12} & \cdots & s_{1n} \\ \vdots & & & \vdots \\ s_{n1} & \cdots & & s_{nm} \end{bmatrix} \begin{bmatrix} I_1 \\ \vdots \\ I_n \end{bmatrix} \quad (B.1)$$

In a theoretical ideal device, certain ports are isolated from each other. For example, in the quadrature hybrid junction shown below, Port 1 is isolated from Port 2, and 3 from 4. Therefore, these elements in the matrix are

$$s_{12} = s_{21} = s_{34} = s_{43} = 0 \quad (B.2)$$

If all ports are matched (VSWR = 1 or $\Gamma = 0$), the diagonal elements equal zero (diagonal elements are reflection coefficients). Off-diagonal elements are the transmission coefficients.

$$s_{11} = s_{22} = s_{33} = s_{44} = 0 \quad (B.3)$$

The junction is reciprocal, making the elements of the matrix symmetrical about the diagonal. Therefore,

$$s_{12} = s_{21}, s_{31} = s_{13}, s_{41} = s_{14}, s_{42} = s_{24}, s_{43} = s_{34}, s_{32} = s_{23} \quad (B.4)$$

The quadrature hybrid junction and scattering matrix shown below reflect the comments made above.

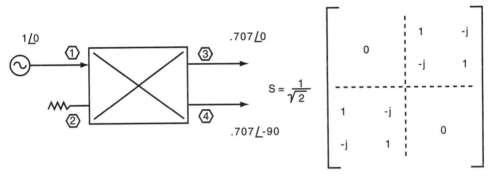

Figure B-1 Quadrature hybrid network for producing equal signals, but with a 90-degree phase differential.

If we perform the matrix operation, we obtain:[1]

$$\begin{matrix} \text{Output} & & \text{Input} \end{matrix}$$

$$\begin{bmatrix} 0_1 \\ 0_2 \\ 0_3 \\ 0_4 \end{bmatrix} = (1/\sqrt{2}) \begin{bmatrix} 0 & 0 & 1 & -j \\ 0 & 0 & -j & 1 \\ 1 & -j & 0 & 0 \\ -j & 1 & 0 & 0 \end{bmatrix} \begin{bmatrix} I_1 \\ I_2 \\ I_3 \\ I_4 \end{bmatrix} \quad (B.5)$$

---

1. The product of the two matrices is defined only when the number of columns of matrix S equals the number of rows of matrix I, a condition satisfied above. Therefore, we obtain

$$0_1 = (1/\sqrt{2})(I_3 - jI_4)$$
$$0_2 = (1/\sqrt{2}) - j(I_3 + I_4)$$
$$0_3 = (1/\sqrt{2})(I_1 - jI_2)$$
$$0_4 = (1/\sqrt{2})(-jI_1 + I_2)$$

but $I_2 = I_3 = I_4 = 0$

$\therefore 0_1 = 0$

$0_2 = 0$

$0_3 = I_1/\sqrt{2} = (I_1/\sqrt{2})\angle 0°$

$0_4 = -jI_1/\sqrt{2} = I_1\angle -90/\sqrt{2} = (I_1/\sqrt{2})\exp(-\pi/2)$

Two equal outputs from ports 3 and 4, but in phase quadrature.

App. B • Scattering Matrix of the Quadrature Hybrid

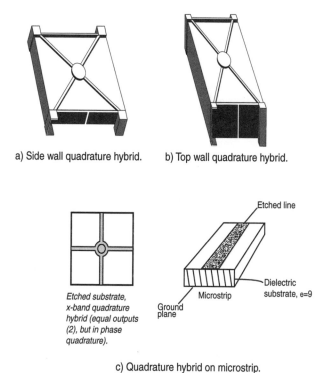

a) Side wall quadrature hybrid.

b) Top wall quadrature hybrid.

Etched substrate, x-band quadrature hybrid (equal outputs (2), but in phase quadrature).

c) Quadrature hybrid on microstrip.

Figure B–2 X-band quadrature hybrid junctions (all about full scale).

APPENDIX C

Example of Trunking and Erlang Tables

C.3 Examples

- 1 voice circuit in use 100% of the time = 1 Erlang
- 2 voice circuits in use 50% of the time = 1 Erlang
- 4 voice circuits in use 50% of the time = 2 Erlangs
- 10 voice circuits in use 10% of the time + 5 voice circuits in use 45% of the time = 3.25 Erlangs (i.e., $10 \times 0.1 = 1\text{E}$ and $5 \times 0.45 = \frac{2.25\text{E}}{3.25\text{E}}$)

The Erlang-B formula determines the probability that a call is blocked.

$$\text{prob(blocking)} = \frac{A^C}{C! \sum_{k=0}^{C} (A^k/k!)} \quad (C.1)$$

where C is the number of channels and A is total traffic intensity offered in Erlangs.

Table C–1 Sample, Blocked-Calls-Cleared Table (Erlang-B)

Offered Traffic [Erlangs]

| N | \multicolumn{12}{c}{Probability of Blocking} |

N	1%	1.2%	1.5%	2%	3%	5%	7%	10%	15%	20%	30%	40%	50%
1	0.101	0.0121	0.0152	0.020	0.0309	0.0526	0.0753	0.111	0.176	0.250	0.429	0.667	1.00
2	0.153	0.168	0.190	0.223	0.282	0.381	0.470	0.595	0.796	1.00	1.44	2.00	2.73
3	0.455	0.489	0.535	0.602	0.715	0.899	1.05	1.27	1.60	1.92	2.63	3.47	4.59
4	0.869	0.922	0.992	1.09	1.25	1.52	1.74	2.04	2.50	2.94	3.89	5.02	6.50
5	1.36	1.43	1.52	1.65	1.87	2.21	2.50	2.88	3.45	4.01	5.18	6.59	8.43
6	1.90	1.99	2.11	2.27	2.54	2.96	3.30	3.75	4.44	5.10	6.51	8.19	10.3
7	2.50	2.60	2.74	2.93	3.24	3.73	4.13	4.66	5.46	6.23	7.85	9.79	12.3
8	3.12	3.24	3.40	3.62	3.98	4.54	4.99	5.59	6.49	7.36	9.21	11.4	14.3
9	3.78	3.91	4.09	4.34	4.74	5.37	5.87	6.54	7.55	8.52	10.5	13.0	16.2
10	4.46	4.61	4.80	5.08	5.52	6.21	6.77	7.51	8.61	9.68	11.9	14.6	18.2
11	5.15	5.32	5.53	5.84	6.32	7.07	7.68	8.48	9.69	10.8	13.3	16.3	20.2
12	5.87	6.05	6.28	6.61	7.14	7.94	8.60	9.47	10.7	12.0	14.7	17.9	22.2
13	6.60	6.79	7.04	7.40	7.96	8.83	9.54	10.4	11.8	13.2	16.1	19.5	24.2
14	7.35	7.55	7.82	8.20	8.80	9.72	10.4	11.4	12.9	14.4	17.5	21.2	26.2
15	8.10	8.32	8.61	9.00	9.40	10.6	11.4	12.4	14.0	15.6	18.8	22.8	28.2
16	8.87	9.10	9.40	9.82	10.5	11.5	12.3	13.5	15.1	16.8	20.2	24.5	30.1
17	9.65	9.89	10.2	10.6	11.3	12.4	13.3	14.5	16.2	18.0	21.7	26.1	32.1
18	10.4	10.6	11.0	11.4	12.2	13.3	14.3	15.5	17.4	19.2	23.1	27.8	34.1
19	11.2	11.4	11.8	12.3	13.1	14.3	15.2	16.5	18.5	20.4	24.5	29.4	36.1
20	12.0	12.3	12.6	13.1	13.9	15.2	16.2	17.6	19.6	21.6	25.9	31.1	38.1
21	12.8	13.1	13.5	14.0	14.8	16.1	17.2	18.6	20.7	22.8	27.3	32.8	40.1
22	13.6	13.9	14.3	14.8	15.7	17.1	18.2	19.6	21.9	24.0	28.7	34.4	42.1
23	14.4	14.7	15.1	15.7	16.6	18.0	19.2	20.7	23.0	25.2	30.1	36.1	44.1
24	15.2	15.6	16.0	16.6	17.5	19.0	20.2	21.7	24.1	26.4	31.5	37.7	46.1
25	16.1	16.4	16.8	17.5	18.4	19.9	21.2	22.8	25.2	27.7	32.9	39.4	48.1
26	16.9	17.3	17.7	18.3	19.3	20.9	22.2	23.8	26.4	28.9	34.3	41.0	50.1
27	17.7	18.1	18.6	19.2	20.3	21.9	23.2	24.9	27.5	30.1	35.7	42.7	52.1
28	18.6	19.0	19.4	20.1	21.2	22.8	24.2	25.9	28.7	31.3	37.2	44.4	54.1
29	19.4	19.8	20.3	21.0	22.1	23.8	25.2	27.0	29.8	32.6	38.6	46.0	56.1
30	20.3	20.7	21.2	21.9	23.0	24.8	26.2	28.1	30.9	33.8	40.0	47.7	58.1
31	21.1	21.5	22.1	22.8	23.9	25.7	27.2	29.1	32.1	35.0	41.4	49.3	60.1
32	22.0	22.4	22.9	23.7	24.9	26.7	28.2	30.2	33.2	36.2	42.8	51.0	62.1
33	22.9	23.3	23.8	24.6	25.8	27.7	29.2	31.3	34.4	37.5	44.3	52.7	64.1
34	23.7	24.1	24.7	25.5	26.7	28.6	30.2	32.3	35.5	38.7	45.7	54.3	66.1
35	24.6	25.0	25.6	26.4	27.7	29.6	31.2	33.4	36.7	39.9	47.1	56.0	68.0
36	25.5	25.9	26.5	27.3	28.6	30.6	32.3	34.5	37.8	41.2	48.5	57.7'	70.0
37	26.3	26.8	27.4	28.2	29.5	31.6	33.3	35.5	39.0	42.4	49.9	59.3	72.0
38	27.2	27.7	28.3	29.1	30.5	32.6	34.3	36.6	40.1	43.6	51.3	61.0	74.0
39	28.1	28.6	29.2	30.0	31.4	33.6	35.3	37.7	41.3	44.9	52.8	62.6	76.0
40	29.0	29.4	30.1	30.9	32.4	34.5	36.3	38.7	42.4	46.1	54.2	64.3	78.0
41	29.8	30.3	31.0	31.9	33.3	35.5	37.4	39.8	43.6	47.3	55.6	66.0	80.0
42	30.7	31.2	31.9	32.8	34.3	36.5	38.4	40.9	44.7	48.6	57.0	67.6	82.0
43	31.6	32.1	32.8	33.7	35.2	37.5	39.4	42.0	45.9	49.8	58.5	69.3	84.0
44	32.5	33.0	33.7	34.6	36.2	38.5	40.5	43.0	47.0	51.0	59.9	71.0	86.0
45	33.4	33.9	34.6	35.6	37.1	39.5	41.5	44.1	48.2	52.3	61.3	72.6	88.0
46	34.3	34.8	35.5	36.5	38.1	40.5	42.5	45.2	49.4	53.5	62.7	74.3	90.0
47	35.2	35.7	36.4	37.4	39.0	41.5	43.5	46.3	50.5	54.7	64.1	75.9	92.0
48	36.1	36.6	37.3	38.3	40.0	42.5	44.6	47.4	51.7	56.0	65.6	77.6	94.0
49	37.0	37.5	38.2	39.3	40.9	43.5	45.6	48.4	52.8	57.2	67.0	79.3	96.0
50	37.9	38.4	39.2	40.2	41.9	44.5	46.6	49.5	54.0	58.5	68.4	80.9	98.0

APPENDIX D

Glossary of Terms

ACG   **Asymptotic Coding Gain.** $G = 20 \log_{10}(d_{\text{free}} / P_{\text{ave}}) / (d_{\text{ref}} / P_{\text{ave}})$ for high signal-to-noise ratio the above becomes
$$G = (E_b / N_o)_{\text{uncoded}} - (E_b / N_o)_{\text{coded}} \text{ dB}$$

ACI   **Adjacent Channel Interference**

ADC   **American Digital Cellular**

AMPS   **Analog Mobile Phone Service**

ARQ   **Automatic Repeat Request** to correct transmissions—as opposed to Forward Error Correction (FEC) where now repeat is required.

ASK   **Amplitude Shift Keying** (one-dimensional equally spaced points signal-space constellation).

Autocorrelation   Degree of similarity between a sequence and a time shifted replica of itself. Pseudo-random sequences have a strong correlation at zero shift point and weak correlation for shifts other than zero.

AWGN   **Additive White Gaussian Noise**

BCH   A type of block coding of data words after the inventors **Bose, Chaudhuri,** and **Hocquenghem.**

BER   **Bit Error Rate** (in most cases expressed as probability of bit error $p(e)$.

BFSK   **Binary Frequency Shift Keying.** No carrier recovery needed: no coherent receiver; two frequencies represent 0,1; frequencies are orthogonal (separated by bit rate $1/T$); bandwidth efficiency, 1 bps/Hz.

BPSK   **Binary Phase Shift Keying.** Phase of transmitted carrier changes by 180° whenever the logic value of the binary data changes; constant envelope signal therefore can perform well in a non-linear channel.

BTA   **Basic Trading Areas** (493).

CCI   **Co-Channel Interference**

CCITT   **International Telegraph & Telephone Consultative Committee**

CDMA   **Code Division Multiple Access.** Several spread spectrum systems accessing a transponder in a satellite (for example) simultaneously

CELP   **Code Excited Predicted Codec** (used in IS-95).

Chip   In a spread spectrum system, a chip is a simple element of the spreading code, that is, the duration of one pulse.

Chip rate   The number of chips (pulses) or code elements per second generated by the spreading code generator.

Clock   Clock pulses synchronize the timing of the shift register used in generating a spread spectrum code sequence. A shift register is an electronic circuit of delay elements which can store one binary digit each. These delay elements, called shift register stages, are connected in series. When these stages receive a clock pulse, the binary digit stored in each stage is shifted forward to the next stage. A simple shift register is shown below. Each stage is a flip-flop circuit.

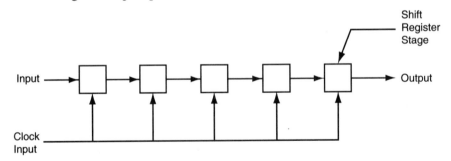

Code division multiplex   A spread spectrum system in which each signal uses a unique spreading code. These codes should be orthogonal to each other to avoid mutual interference.

CODE RATE   The ratio per unit time of the *message* bits to total number of bits transmitted.

CODEC   **(coder-decoder)** Tandem A/D converter; digital precessor; D/A converter.

CODING GAIN The reduction in E_b/N_o required for providing a specified bit error rate (BER).

CPFSK Continuous Phase Frequency Shift Keying

CPM **Continuous Phase Modulation.** Basically a frequency modulation in which the baseband signal is a digital signal instead of a familiar signal such as voice, music, or TV; phase is continuous at symbol boundaries; no jump-in phase like in PSK; reduces spectral sidelobes and thus ACI and ISI; constant envelope; MSK, GMSK, and TFM in this family.

Cross-correlation Degree of similarity between two different sequences. Two different orthogonal sequences have zero correlation between them. The fewer the prime factors in the sequence, the less the cross-correlation. For example, for $n = 5$, sequence length is 31 and there is only one prime factor. In this case n has been referred to as a primitive, or Mersenne prime.

CT-2 **Cordless Telephone** (2nd generation digital).

d MINIMUM Defined as the minimum Euclidean distance between signal points in signal space constellation, for *uncoded* signal. Sometimes referred to as the reference distance, d_{ref}.

D-AMPS **Digital AMPS** (IS-54)

DECT **Digital European Cordless Telephony**

Δf **Frequency deviation of digital FM waveform**; equal to $f = 2\pi h g(t)$, where h is the modulation index and $g(t)$ is the output of the shaping filter (if used).

DPSK **Differential PSK.** Information is converged in the phase difference of a carrier signal between consecutive symbols.

D/R **Co-channel cell distance to cell radius ratio**

DSC **Digital Communication System**

DSI **Digital Speech Interpolation**, releases the transceiver channel during the silent periods of speech, vertically doubling the number of channels.

DVSB Digital Vestigial Sideband (note this is comparable to PAM where there are M-ary discrete amplitude levels.)[1]

$E_b \triangleq$ **Energy per Bit**

---

1. The proposed HDTV standard will use 8-VSB and 16-VSB modulations.

$E_s \triangleq$ **Energy per Symbol**
$= E_b \log_2 M = E_b (\log_{10} M) / \log_{10} 2$

ED   **Euclidean distance**

EIRP   **Effective Isotropic Radiated Power**

Error event   Identified as a departure from the all-zeros path followed by a remerging with the all zeros path.

E-TDM   **Extended TDMA**, Hughes system providing six users per channel; use DSI.

Euclidean distance   Distance between signal points in the signal space constellation.

Excess phase   Signal phase during a bit interval, which is above the linear phase provided by the carrier.

FDMA   **Frequency Division Multiple Access**; used in AMPS.

FEC   **Forward Error Correction**; detect and correct errors at receiver without repeat transmissions. Two types of FEC: block coding and convolutional coding.

4-PSK   Same as QPSK

Free euclidean distance, d_{free}   Defined as the minimum Euclidean distance between any pair of transmitted code sequences (which split at some level in the trellis and merge later. The free Euclidean distance between any two code sequences is defined as the square root of the sums of the squares of the geometric distances between corresponding symbols of the two sequences.

FSK   **Frequency Shift Keying**

Full response CPM   The phase response is shaped over a full symbol period, T.

GMSK   **Gaussian MSK**. CPFSK with Gaussian-shaped pre-FM modulation filter; substantially suppresses out-of-band spectrum (function of LPF bandwidth), thus permitting reduced channel spacing; a modest cost in BER; used in European cellular GSM; constant envelope.

Gold codes   Codes manifesting bounded cross-correlation functions. Can be generated by modulo-2 addition of two preferred maximal-length sequences, but they themselves generally are not maximal-length sequences. Used in CDMA systems since they are quasi-orthogonal, but bounded. Generally maximal-length sequences are not suitable for CDMA.

Gray code Single bit difference from one signal symbol to another. For example, 000, 010, 011, 111.

GSM Originally named **Groupe Speciale Mobile**, but later changed to **Global System for Communications**. Eurpoean cellular standard which uses GMSK modulation with BT = 0.3; appears to be becoming a global defacto standard.

h   symbol for modulation index in digital FM. Usually equal to h = 1/2; analgous to β in analog FM.

Hamming distance The number of places any two codes differ. For example, 00010 and 11001, $D = 4$. The greater the Hamming distance of a code, the greater is its ability to correct a corrupted message. A larger Hamming distance is achieved by using a more complex code. Note in TCM we deal with Euclidean distances as opposed to Hamming distances.

Hamming weights Defined to be the number of non-zero components of the code. For example, 011001, $W = 3$.

Interference margin The amount of interference a system is capable of withstanding while still providing necessary *S/N* ratio for reliable reception. This is given as
$$M_I = PG - L - (S/N)_{req}$$
where *PG* is the processing gain, *L* is the implementation loss, and $(S/N)_{req}$ is the required output *S/N* ratio after demodulation (this is degraded by interference). The larger the ratio of the BW of the spread signal to that of the information signal, the smaller the effect of unwanted signal interference.

IOC **Intermediate Orbit Communications**. Also referred to as medium earth orbit (MEO).

ISI **Intersymbol Interference**. Interference between adjacent symbols.

IS-54 U.S. standard, **DAMPS**; TDMA uses three channels in one AMPS channel (30 kHz).

IS-95 **CDMA with QPSK/BPSK modulation** (note CDMA is not the modulation).

JDC **Japanese Digital Cellular**

K   **Encoder constraint length in convolutional encoding.** The higher the *K*, the greater the coding gain, but with circuit complexity. Values typically less than *K* = 10, but *K* up to 31 has been used in space applicaitons.

LIP   **Low probability of intercept.** A term indicating that a signal is difficult to detect. Spread spectrum signals fall into this category. For spread spectrum, a signal may be deep in noise.

M   **Size of alphabet (or number of states),** e.g., 4-PSK = 4 output phases, *M* = 4.

M-ary PSK   **Multi-point Quadrature Amplitude Modulation (M-PSK).** Usually rectangular constellation (need not be) in signal phase space; non-constant envelope signal; penalized when transmitted through a non-linear channel; good bandwidth effeciencies, but not power efficient.

Maximal-length sequence   Longest sequence that can be generated by a given length, *n*, stage shift register.

Mersenne prime   Linear maximal-length sequences that have code lengths equal to a prime number. Example: 3,7,31,127, ... (the fewer the prime factors in a sequence, the smaller the cross-correlation values).

MIC   **Microwave Integrated Circuit**

MMIC   **Monolithic Microwave Integrated Circuit**

MODULO-2 Addition   Designated by symbol $\oplus$. Arithmetic addition of binary digits. Examples of binary addition are:

$$0 \oplus 0 = 0$$
$$0 \oplus 1 = 1$$
$$1 \oplus 0 = 1$$
$$1 \oplus 1 = 0$$

M-PAM   **M-ary Pulse Amplitude Modulation** (one dimensional).

M-QAM   **M-ary QAM.**

MSC   **Mobile Switching Center**

MSK   **Minimum Shift Keying.** FM with modulation index of 0.5; constant envelope; similar to staggered QPSK with half sinusoidal filter weighting; phase continuity in the RF carrier at the bit transitions; less spectrally efficient than BPSK, QPSK, and OQPSK; minimum speraration between binary frequencies (separated by bit rate, 1 / *T*); manifests out-of-bounds radiation; spectral roll-off $1/f^4$; sometimes referred to as Fast FSK (FFSK).

MTA **Major Trading Areas** (51).

MTS **Mobile Telephone Service** (single cell, high power, percursor to cellular concept).

MTSO **Mobile Telephone Switching Office**

$n = \log_2 M$ **Number of bits per symbol**, e.g. $M = 4$, $n = \log_{10} 4 / \log_{10} 2 = (0.6/0.3) = 2$

Near-far performance A nonrelated user which is close to a receiver will interfere with a bona fide signal which is matched to that receiver, but farther away. This problem is alleviated by automatic power control.

NLA **Non-Linear Amplifier** (saturated Class-C amplifier). More RF-power efficient than linear amplifiers; lower cost; requires less battery power than linear amplifiers.

N_o **Noise power density measured in W/Hz**

OQPSK **Offset QPSK.** More robust against degradation in filtering and non-linear processing than QPSK (no 180° phase changes); same spectral roll-off as QPSK; no spectral restoration after bandwidth limiting; sometimes referred to as staggered QPSK.

Orthogonal When two sequences correlated (cross-correlated) with respect to each other are zero, the sequences are said to be orthogonal. Codes that have cross-correlation coefficients (number of bit agreements minus bit disagreements) equal to zero for all pairs in the set of codes are also orthogonal.

    e.g., code 1 : 101101
        2 : 110001

The first, fifth, and sixth bits are in agreement = 3

The second, third, and fourth are in disagreement = 3

Therefore, since the number of agreements – number of disagreements = 0, the codes are orthogonal and their cross-correlation is equal to zero.

PAM **Pulse Amplitude Modulation**

Parallel transitions (in a trellis) Results from the transmission of uncoded bits along with coded bits in the convolutional encoder.

Partial response CPM Only part of the symbol shaping is done over a symbol period, T.

PCN   **Personal Communication Network**

PCS   **Personal Communication Service**

π/4-QPSK **Quadrature PSK.** No ±180° transitions as in PSK (good), but ±45°, ±135°; non-constant envelope; used in Americn IS-54 and Japanese PDC systems (cellular). Requires linear amplification.

PN   **Pseudo-noise.** A signal that appears noise-like, but is generated by deterministic means (shift register) and is repetitive.

Processing gain In spread spectrum, the ratio of the RF bandwidth of the spread signal to the information bit rate.

$$PG = 10 \log_{10}(BW_{RF}/\text{data rate}) \text{ dB}$$

PSK   **Phase Shift Keying**

PSTN   **Public Switch Telephone Network**

QAM   **Quadrature Amplitude Modulation (two-dimensional).** A modulation where information is transmitted in the amplitude of cosine and sine components of the carrier.

QASK   **Quadrature Amplitude Shift Keying** (same as QAM).

QPSK   **Quaternary PSK.** Very popular in terrestrial and satellite communications; constant envelope, thus suitable for use in non-linear channels; if bandwidth limited, no longer constant envelope (limiting restores constant envelope, but also restores sidelobes); phase changes can be 0°, ±90°, or 180°; spectral roll-off $1/f^2$.

r   **Coding rate**

Rake receiver Spread spectrum demodulator with several correlators (sometimes called "fingers"), each synchronized to different multipath components of the received signal. Combines delayed versions of the signal, mitigates delay spread problem, and provides diversity gain (time).

Random sequence A sequence that is completely random; an unpredictable sequence (e.g., flip of an unbiased coin).

R-S   **(Reed-Solomon)** Block code for error detection and correction. Effective against bursty errors.

SMSK **Serial MSK.** Constant envelope; less degradation by carrier phase error than SQPSK; serial processing rather than parallel processing; used in NASA's ACTS satellite.

Soft Decision (as opposed to hard decision (yes/no)) A soft decision demodulator first decides whether the output voltage is above or below threshold and then computes a confidence number how far from threshold it is. Soft decision in the decoder, in lieu of hard decision, realizes about a 2 dB improvement in coding gain. Used with the Viterbi decoder in TCM.

Spreading code Binary sequence used to spread the spectrum of direct sequence signals. Its rate is several times greater than the baseband (intelligence) signal.

SRG **Shift register generator.** A sequence generator that uses a shift register and modulo-2 adder to generate a pseudo-random sequence.

SSMA **Spread Spectrum Multiple Access**

symbol rate bit rate $R_s = R_b / \log_2 M$

Symbols Member of the M-ary alphabet that is transmitted during each symbol duration, T_s. One M symbol is transmitted during each symbol duration. One or more bits can be sent as a symbol. A symbol may be sent as a digital *level*.

TACS **Total Access Communications** (British).

TASI **Time Assignment Speech Interpolation**

TCM **Trellis Coded Modulation.** (implies that convolutional is used as opposed to block coding, one can implement TCM-like with block coding)

TDMA **Time Division Multiple Access**

TFM **Tamed Frequency Modulation.** Form of CPM where phase changes occur over several symbols. The spectrum is comparable to GMSK with a BT = 0.2; constant envelope.

VCO **Voltage Central Oscillator**

VLSI **Very Large Scale Integration**

VSELP **Vector Sum Excited Linear Prediction** (used in IS-54 voice CODEC).

Index

Numerics

256-QAM TCM cross constellation 118

A

ACI 34
Adaptive array 241, 242
Adaptive array in a quiescent signal field 234
Allocated 164
Alternate method of generating an MSK 47
AMPS 2
Applebaum 244, 246, 257, 258, 259
ARQ 70
AWGN 2

B

Bandlimited QPSK 65
Bandwidth efficiency 15, 18
Bandwidth/power efficiency 7
Basic difference 257
Benefits from using CDMA cellular 185
BER 69, 70
Berlekamp-Massey 71

Block code (n, k, t) generation 84
Block codes 70
Block coding 82
Bose, Chaudhuri, Hocquenghem (BCH) 71
Bursty-type errors 80
Butler matrix 193, 198
Butler matrix beam-forming network 193
Butler matrix labyrinth 219

C

Cancellation beam 250
Cataloging-in-Publication Data ii
CDMA 128
Cell splitting 175
Cellular standards 179
Channel capacity 5
Class-A amplifiers 35, 36
Class-C amplifiers 36
Code generation 133
Code rate 71
Codex 2660 118
Codex 2680 118
Coding gain 75, 80
Coding tree 94
Cofactor 237
Coherent combination of signals 227
Coherent noise 232
Column vector 243
Complex weight 241
Concatenated codes 74
Concatenated coding 98
Conjugate phase shifts 241
Constellation of 256 points 118
Constellations 15, 22
Constraint length, K 71, 75, 91
Continuous phase modulation 33, 42
Convolutional codes 70
Correlation 61
Correlation mixer 204

Index

Covariance matrix 244
Cusps 50

D

Data rate 21
Delay spread 2, 260
Diagonal matrix 246
Digital implementation of a switched-beam antenna 225
Diplexers 223
Direct sequence 130
Duplexer 224

E

Element factor 221
Error control coding 69
Euclidean distance 19
Excess phase 50
Eye diagram 23

F

Fast frequency shift keying (FFSK) 44
FM processing gain 129
FM threshold criterion 2
Forward error correction 70
Free Euclidean distance 107, 111
Frequency hopping 130
Full response 48

G

Gaussian LPF 53
Generation of Gaussian MSK 53
Generation of minimum shift keying (MSK) 43
GMSK Modulation 51
Golay 71
Gold codes 141
Gray coding 20
Gray encoding 21

H

h 46
Hamming 71, 80, 83, 85
Hamming distance 89, 94
Handover 164
Hard decision 75, 76, 77
Higher order modulation methods 15

I

I 37
I is an identity matrix 246
Impulse response of Gaussian filters 54
Increasing capacity 174
Information rate 5
Interference cross-correlation noise 147
Interleaving 81, 98, 99, 101
Inverse 236, 247
I_o 146, 149
ISBN ii
ISI 2

L

LEO 1
Linear modulation 34
Linkabit 77

M

Mean reuse distance 210
MEO 1
Mersenne primes 145
MMIC 2
Modulation 9
Modulation index of h = 0.5 44
MSC 163, 164
MSK 42, 44, 45, 65
Multipath 152
Multipath phenomenon 152

Index

Multiple (5) adaptive sidelobe cancellations 206
Multiple volumetric beams 197

N

NASA data 100
NRZ 36

O

ODYSSEY 193
Offset QPSK (OQPSK) 41
OQPSK 45
Orthogonal beams 200, 220

P

Partial response 48
PCS Standards (1900 MHz) 187
Perturbations 17
Phase behavior 59
Phase trellis 48, 50
PN 135
PN Code Gen. 130
PN Seq. Gen. 130
PN sequence 136
Power spectra of GMSK 55
Practical power spectra for GMSK 56
PRC 131
Processing gain 148
Pseudo-random sequence 136
PSTN 163, 164

Q

Q 37
QPSK 41, 45
Quiescent pattern 249

R

RAKE 154, 178
RAKEing 153
Reed-Solomon 71, 81, 88
Regrowth 40
Resting locations 50, 51
Roaming 185
Rosette of directional beams 216
Row vector 243

S

Sectorization 211, 212
Sequential 78
Sequential decoding 82
Shannon 5
Shannon bound 7, 12, 19, 70
Sidelobe cancellers 203
Signal orthogonality 59
Signal state space diagrams 17
SINR 241
Skybridge 1
Soft decision 76, 77, 78
Space channel 100
Spectral regrowth 36
Spread spectrum 127
Steering vector 246, 247
Switched-beam smart antennas 209
Symbol 19

T

Tamed frequency modulation 58
TCM 97
Technology maturation 177
Teledesic 1
Terrestrial-based wireless communications 161
Tessellated 162
TFM 59
Time diversity 154

Trade-off performances 11
Transpose operation 243
Trellis 57
Trellis coded modulation 17, 105
Trellis representation of the code tree 96

U

UHF 164
Ungerboeck 97, 107, 111

V

VCO 47
Virtual location 52
Virtual receive beam 244
Viterbi 71, 92, 94
VLSI 2, 119
Voice activation 151

W

Widrow 236, 256, 257, 258, 259, 260

The Author

Bruno Pattan has a diversity of experience with his tenure in engineering extending to 35 years. Approximately half this period was as a radar systems/electronic warfare engineer. Studies in this area included coherent radar system design, radar detection, pulse compression, and phase array design for ICBM, tactical, and airborne radar applications. Allied to this work was the evaluation of enemy radar sensors and missile systems, as well as the assessment of countermeasures against these threats. Additional industry experience includes satellite system studies addressing threats to communications satellites (MILSATCOMS) and satellite system design central to antennas/microwave subsystems.

After joining the government, he became a principle investigator on several programs such as direct broadcast satellites, Non-GSO mobile satellites, and CCIR satellites. More recently he has performed studies on cellular/PCS technology, including digital modulation methods, and mobile antennas (classical and Phased arrays). The latter also included smart antennas.

He has given talks at satellite conferences and has contributed material to several publications. He is the author of two books on satellites and satellite-based cellular. He is a member of the AIAA and IEEE, for which he also reviewed papers. He has served on several U.S. government committees and has been a U.S. delegate in Geneva for meetings on satellite matters.

Before accepting a position with the FCC's Office of Engineering and Technology as a technical analyst, his previous affiliations were with GTE, Lockheed Electronics, United Technologies, and Computer Sciences Corporation.

PRENTICE HALL
Professional Technical Reference
Tomorrow's Solutions for Today's Professionals.

Keep Up-to-Date with
PH PTR Online!

We strive to stay on the cutting-edge of what's happening in professional computer science and engineering. Here's a bit of what you'll find when you stop by **www.phptr.com**:

- @ **Special interest areas** offering our latest books, book series, software, features of the month, related links and other useful information to help you get the job done.

- ☞ **Deals, deals, deals!** Come to our promotions section for the latest bargains offered to you exclusively from our retailers.

- $ **Need to find a bookstore?** Chances are, there's a bookseller near you that carries a broad selection of PTR titles. Locate a Magnet bookstore near you at www.phptr.com.

- ! **What's New at PH PTR?** We don't just publish books for the professional community, we're a part of it. Check out our convention schedule, join an author chat, get the latest reviews and press releases on topics of interest to you.

- ✉ **Subscribe Today!** **Join PH PTR's monthly email newsletter!**

  Want to be kept up-to-date on your area of interest? Choose a targeted category on our website, and we'll keep you informed of the latest PH PTR products, author events, reviews and conferences in your interest area.

  Visit our mailroom to subscribe today! **http://www.phptr.com/mail_lists**